AGRICULTURAL PROJECT MANAGEMENT

Monitoring and Control of Implementation

PETER SMITH

Honorary Research Fellow, Centre for Agricultural Strategy, Reading University, UK (formerly Resident Adviser to World Bank projects in Sind Province, Pakistan)

ELSEVIER APPLIED SCIENCE PUBLISHERS
LONDON and NEW YORK

ELSEVIER APPLIED SCIENCE PUBLISHERS LTD
Ripple Road, Barking, Essex, England

Sole Distributor in the USA and Canada
ELSEVIER SCIENCE PUBLISHING CO., INC.
52 Vanderbilt Avenue, New York, NY 10017, USA

British Library Cataloguing in Publication Data

Smith, Peter.
    Agricultural project management.
    1. Agricultural development projects—
    Developing countries—Management
    I. Title
    630′.68′4

ISBN-13:978-94-011-5935-7   e-ISBN-13: 978-94-011-5933-3
DOI: 10.1007/978-94-011-5933-3

WITH 24 TABLES AND 36 ILLUSTRATIONS

© ELSEVIER APPLIED SCIENCE PUBLISHERS LTD 1984

Softcover reprint of the hardcover 1st edition 1984

Photoset in Malta by Interprint Limited

AGRICULTURAL PROJECT MANAGEMENT

Monitoring and Control of Implementation

Preface

I first became interested in the methods of planning the sequence and timing of jobs on large-scale development projects, as a field officer involved in planning and implementing mechanised farming schemes in Uganda in the mid-sixties. This interest was reinforced by experience of agro-industrial projects in both Nigeria and Iran, when it became obvious that the lax traditional methods of both planning and controlling the implementation of agricultural and other rural development projects were very ineffective compared with those already in use in other disciplines. An extended spell as Resident Adviser on a World Bank project to strengthen planning and project management services in the agricultural sector in Sind Province, Pakistan, stimulated this interest further, and gave opportunities to develop the use of improved methods on some very complex schemes.

This book summarises the experience gained in adapting critical path methods, well established in other fields, to Third World development projects, with their peculiar problems.

It would not have been possible to reach this point without the help and stimulation of discussions with a large number of colleagues, including John Joyce (then of Hunting Technical Services), Hatsuya Azumi (World Bank), and—particularly—Zaffar Sohrwardy and Akhtar Ali of Aarkays Associates in Karachi, during our work together.

My thanks are also due to Yasin Mohammed, who typed most of the original draft; Anwar Mohammed and Irene Mills for final typing; and to my wife, Jill, for drawing the original figures.

PETER SMITH
High Greenleycleugh,
Ninebanks,
Hexham,
Northumberland,
UK

Contents

CHAPTER 1

The Scope of Project Management

1.1. INTRODUCTION

In our context—agricultural development—a project is a unique operation to modify the means of production. This definition includes a huge range of activities, from an individual farmer constructing a large new dairy unit, to a government department executing large scale conservation works, or even to a project in which the major items are administrative rather than physical, such as the reorganisation of part of the state's agricultural services.

The boundaries of a project will depend on the interests and responsibilities of the agency involved: for example, for the sponsor of the scheme—usually a government department—the project consists of everything required to bring the scheme into production. However, for a civil engineering contractor on the operation, his contractual obligations alone constitute *his* project.

Each of these agencies has to manage the physical activities and flow of funds within its area of responsibility. It is in attempting to do this that the big difference between projects and production (using existing capital assets) is revealed: many of the management tools and techniques which are so useful to the production manager rely on observing regular practice. Work study and activity sampling do this with a view to improving the standard methods; and management by exception uses deviations from the regular pattern as a signal that intervention is necessary. Regular practice, in production management, is of massive importance as a bench mark. This reference point is usually not available to the project manager: we have, by including the word 'unique' in the definition of projects, separated off a class of undertakings which are essentially one-off jobs, on which there is no regular practice to refer to.

1

No two major irrigation projects, no two large-scale range improvement schemes, will be sufficiently alike to enable the first to serve as a detailed guide on how the second should be implemented, despite the fact that some of the components may be identical. Therefore, project management requires a distinct set of techniques to collate and interpret information relevant to controlling the scheduling of activities and the flow of funds in a way that provides a reliable alternative reference point. These techniques are relevant in a wide range of situations: in large-scale investments on individual farms in the developed countries, in traditional agricultural development projects in the less developed countries (LDCs), and in the management of the work of an individual contractor on a large project.

Suitable techniques do exist, and have been widely used in a variety of commercial and engineering applications, in many countries; agriculture is one of the few fields in which they have not been applied. The whole question of the effectiveness of project management systems is of particular concern in agricultural projects in the LDCs, and this is the field which we shall mainly be discussing; but the methods and comments, with minor and obvious changes, apply equally well to agricultural projects anywhere.

A second point of difference between project management and most production management is that the project manager has to be a man who coordinates the efforts of technical specialists. Unlike the manager of, say, an engineering factory, who will be dealing with a single discipline, he must be able to integrate the efforts—and interpret the reports—of civil engineers, utilities agencies (e.g. gas or electricity suppliers), builders, equipment installers, and his agricultural staff.

1.1.1. The Need for Project Management
The reasons why agricultural development projects are such a centre of concern are: the urgency of increasing food and fibre production in the poorer countries; the amount of investment that is being applied, by both the recipients and by aid agencies, apparently without proportionate returns. Some of the reasons for this arise outside the sphere of project management: for example, a project may be ill-conceived from the start, because it requires an input of working capital or labour by farmers who are unable or unwilling to make the necessary contributions. However, to suggest that all projects which are significantly less effective than their planners intended were ill-chosen is undoubtedly wrong: many are marred in the making, as a result of disorganised implementation. This

usually results in incomplete implementation of the original plan, or such delays that the investment is near useless, by reason of deterioration of parts of the project. Many sad examples of this have occurred on resettlement projects, where disjointed implementation has resulted in settlers arriving before the promised roads, or schools, or water supplies have been provided, with the result that they become disgruntled and alienated.

1.1.2. Project Management: Relationship with Monitoring and Evaluation
There are sufficient cases in both categories (ill-conceived and badly implemented projects) to have generated a strong interest among those actively involved in agricultural development in the whole area of project control and assessment, which is often referred to as monitoring and evaluation. Unfortunately, it has not always been clearly appreciated that this phrase embraces three functions, distinct in that they require very different techniques and answer very different questions. These functions are:

(i) Monitoring of implementation—this is examining whether or not the project is being implemented according to the current plan. The project management systems described here can fulfil this function.

(ii) Monitoring of operation—this is examining whether or not the completed project is being operated according to the original specifications. For example, in the case of a project to set up a new extension service, monitoring of operation would check whether the required number of visits were being made by extension workers. This function relies heavily on statistical reports; other available techniques include activity sampling, in which the work of the staff is checked directly, in particular periods selected according to a proper statistical plan, to find out what they actually are doing and why.

(iii) Evaluation—investigating whether or not this project, once implemented, has produced the benefits forecast, at the price budgeted for. This question can be tackled by econometric methods. For example, to test the success of a project designed to increase farmers' use of fertiliser by extension efforts and an improved distribution system, the evaluator might compare the trend and fertiliser usage in districts inside the project with the trend in projects outside the district, statistically corrected for other relevant factors which could affect usage, such as availability of irrigation water.

Here, we will be concerned almost exclusively with the systems that form the base of the implementation monitoring pyramid: we are looking in detail at the implementation phase of schemes, rather than their operational phase. Obviously, a government or an aid agency is not concerned with all the details of every one of its projects, but only with whether the major events in the project are happening on time. A monitoring system will be expected to provide reports and projections in these terms. However, unless both the target dates for such events, and the current estimates of event dates are firmly grounded in accurately detailed reality, this part of the monitoring function will fail, and the information that will emerge will be the actual dates of events that have happened, with no objective explanation of any drift from schedule, and no reliable forecast of the dates of events that are yet to happen. Implementation monitoring is impossible without good project management systems—basically because these provide the contact with the underlying physical realities of the work.

Much of what has been said so far is about techniques, and it is important to get the whole question of quantitative techniques in perspective: management is not about mathematical or graphical techniques. The most important function of a project manager is to ensure that all the personnel and organisations involved in the project are cooperating harmoniously, and giving of their best, by ensuring that objectives are clearly understood, and conflicts are resolved promptly and fairly. No system will remedy a situation in which the initiative of staff has been destroyed by abrasive management, and contractors have been alienated by an overly suspicious and inflexible project manager. The techniques to be discussed are tools, no more. At the same time, to attempt to organise a major project without the simple and effective tools that are available is foolish; they are the proper equipment for the job.

1.2. REQUIREMENTS OF AN EFFECTIVE PROJECT MANAGEMENT SYSTEM

A good, effective project management system has to do four things:

— it must control the rate of work and the consequent expenditure;
— it must detect problems (particularly deviations from schedule, but

also the emergence of unforeseen technical problems) early enough
for remedies to be applied;
— it must provide realistic, adequately summarised, and easily in-
terpreted information to its 'owner' on the likely future progress of
the project, both physical and financial;
— it must identify responsibility clearly, both in terms of who should
do particular jobs, and, by implication, who failed to do things on
time. This last item should not be at the top of the list of desirable
features of a project management system, but the knowledge that
incompetence and failure to do the necessary work will be traced
back to those responsible can be a useful incentive. Most project
staff are, after all, only doing a fairly routine job, for an unexciting
salary, which is little encouragement towards a deep commitment
to the project's aims.

To achieve these objectives, three things are needed:

— targets for physical and financial progress;
— a system for giving instructions to staff;
— a channel for feedback of information on what has happened.

These will be dealt with in turn.

(a) *Targets*: The basic set of progress targets must be feasible; that is,
they must contain a sufficient allowance of time for all the items of work
that have to be done. This is a relatively straightforward matter: doing
this is a question of doing for time what budgeting does for funds, in
making sure that everything is correctly allowed for, with no omissions
and no double counting. Nevertheless, it is surprising how often even this
is skimped; it is particularly surprising in the case of contractors for
mechanical and civil engineering items, where physical quantities for
costings are listed in minute detail, often even down to door fittings, and
yet the time taken is estimated by guesswork. As time has an enormous
effect on the contractors' total overheads cost, this is not really very
clever. (Professional economists, are, however, often much worse, and with
less excuse: cost forecasts are made in considerable detail, and abstruse
discounted cash-flow budgets prepared, whose validity is undermined by
a failure to spend sufficient effort on estimating the durations involved—
required as input for those same calculations.)

However, there is a second aspect of feasibility: many activities on a
project interfere with each other, in some way or another, i.e. some
cannot be started before others are completed, and some compete for

resources such as masons, surveyors or bulldozers. When this effect is ignored, project length may be drastically underestimated. In practice, completion times for major projects are often drastically underestimated, and largely because these two factors have been ignored (although the administrator's megalomania is also involved: this is the tendency of senior non-technical administrators to believe that the existence of an order or a minute regarding the finishing date of a project guarantees, not only that it will be finished then, but that it *can* be finished then). The results of such underestimates are many: often, the least damaging effect is the direct effect on costs, to which overheads (including idle staff time and interest on the capital investment tied up in unusable assets for the period of the delay), and cost escalation due to inflation make the largest contributions. Wastage of unused assets can easily be even more expensive: unused items of equipment are pilfered or deteriorate, under-used canals silt up, and uncropped cleared land reverts to bush. But the most damaging effect may be the waste of human assets: farmers become disillusioned—and often contemptuous—of highly publicized projects which, in the field, exist only as a patchwork of uncompleted, unrelated, and useless pieces of work; and staff become disillusioned, with the best of them drifting away, and the others drifting into habits of idleness that will take much breaking. On the financial side, there is a collector's piece of irony: if expenditure targets are over-ambitious because the forecast of physical progress was too optimistic, there will be under-expenditure in the earliest accounting periods. Given the chronic shortage of funds found in most LDCs, this usually results in the allocation for the next accounting period (e.g. financial year) being cut drastically. With only average ill fortune, this cut will neatly castrate the project: just as activity and expenditure are finally beginning to build up, the funds are taken away.

Even now, we have not reached the end of the disasters that can result from the establishment of an unrealistic set of targets. Fairly obviously, at some stage, the targets will have to be compared with actual achievement, to see whether remedial measures are required, either by the project's sponsor, or by any aid agency involved. If the targets are unrealistic, there will be continual false alarms, on the difference between actual and planned progress. However, no-one directly concerned will ever realise they are false, having no means of knowing that the whole set of targets is unachievable. These alarms tend to make the sponsors of the project believe that the project management is incompetent, the project management to feel that it is being unreasonably harassed by the sponsors, and any aid agency involved to believe that the sponsors are either incompetent or out to sabotage the project for mysterious reasons of their own. In this sort of atmosphere, the idea of implementation monitoring is a nonsense, a mere groping in the fogs of self-justification. We have said that targets must be feasible, and looked at what can happen when they are not. There is another criterion to be applied: they must be reasonably close to being optimal, i.e. they should not make unreasonably pessimistic estimates. We will see how this is taken care of later.

(b) *Instructions*: It may seem to be over-emphasising the obvious to include this, but, in practice, the methods by which the targets are communicated to the people who will actually directly supervise the work are often unsatisfactory. They should make sure that the right man is told, in sufficient detail, and at the right time, what he must do. Often, no formal provision is made for this, and junior staff are expected to deduce their detailed responsibilities from a circulated generalised overall description of the project.

(c) *Feedback reports*: As with targets, these have to be grounded in a detailed knowledge of how the project fits together: especially, which activities can hold up which others. Without this, there can be no objective identification of the reasons for delay, and therefore no reliable basis for forecasts of future progress. Equally important, without a proper identification of the reasons for delay, it is very difficult to devise effective remedies. Compared with this, identifying culpable individuals is a relatively trivial problem. Also, the timing of the reporting system must be right: the interval between reports must be sensibly chosen, in relation to the duration of typical jobs, if delays and their causes are to be identified before they become serious.

In addition, a good feedback system will digest the information, eliminating the trivial and concentrating the essential for absorption by the manager and his superiors.

1.3. TRADITIONAL METHODS OF CONTROLLING AGRICULTURAL PROJECTS

1.3.1. An Example
As we have seen in the previous section, a project control system consists of three elements: a set of targets, which may be revised from time to time; a method of converting the targets into detailed instructions, and of

passing these down to the people responsible for the physical execution of the various activities, such as hiring staff, paying bills, digging canals and so on; and a means of obtaining feedback, that is, summarised information on how well the achievement matches the targets.

A variety of loosely-structured systems exists for the control of agricultural (and other non-engineering) projects, and the important features of these are embodied in the following description, which follows fairly closely a system in use in parts of the Indian sub-continent.

In this, the targets are embodied in an official project proposal form. (The primary purpose of this form is to present to central government the information required for making decisions on whether or not to approve the project.) It contains a written description of the undertaking, together with tables of its requirements for physical inputs, staff, and funds in each period of the project's life. (The periods are usually of three

Monthly report

1. Code number of project
2. Name of project
3. Financial statement of _____ 19_____

	Release of funds	Actual expenditure	Expenditure committed (*)	Total
Local cost				
Foreign exchange				
Total				

Signature of Project Director

(*) The committed expenditure should be worked out from purchase orders and contracts (other than those for staff employment) on which payment would be legally due at a later stage.

Fig. 1.1.   Example of a monthly progress report (traditional system).

months duration.) The distribution of requirements over time is usually done 'by common sense' or 'from experience'.

The second element (which should convert the targets into clear, simple instructions, each directed to the man responsible) has, in this system, virtually no formalised existence. All senior staff, that is, those in charge of sections such as land clearance, building construction, and irrigation work, are given a copy of the project proposal form. They interpret it to their subordinates, often on a day-to-day basis. (In at least one case, construction of a large seeds processing plant—costing some three million pounds—was being controlled by the construction contractor's resident engineer and the client's engineer meeting each evening over cups of tea to plan the next day's operations. No longer-term planning was done and no notes were ever kept of the meetings; it is perhaps as well that nothing stronger than tea was available.)

The third element—feedback—is provided by the forms shown in Figs 1.1 and 1.2. Figure 1.1 is a monthly financial summary; Fig. 1.2 is a quarterly summary of physical and financial results. In the latter, there is an item for 'bottleneck codes' which are intended to identify the sources of delays.

1.3.2. Deficiencies of the Example

It is worth emphasising that this system is, in its essentials, actually in use, and has not been constructed with the aim of displaying the largest possible number of faults. Despite this, and despite the fact that large numbers of intelligent government officials have been involved with it, over a long period, there are serious defects in virtually every aspect. The reason for this is basically that it was never designed: it simply grew up this way.

As we review the basic elements—targets, instruction and feedback—in turn, it will also become apparent that the system as a whole does not fit together very well, because its parts were originally designed with other aims in mind: as we have seen, the targets are embodied in a form which was designed to summarise the information needed by central government departments for assessing departmental and provincial development plans. Similarly, the forms providing the feedback are largely the inspiration of departmental accountants, who—quite properly—are concerned with ensuring that expenditure stays within the sanctioned limits, but with little else.

(a) *Targets*: In some ways, the targets do, in fact, represent one of the worst deficiencies: technical data, such as the requirements for tractors,

Quarterly report

Progress report for the quarter ending on _____

1. Code number of project
2. Name of project
3. Financial status:
   (a) Total approved cost of project (from last revision)
   (b) Expenditure up to end of last financial year
   (c) Allocation for current year
   (d) Funds released to the project during current quarter
   (e) Expenditure up to the end of previous quarter
   (f) Expenditure during the current quarter

4. Physical status:

Item of work	Total cost of each item	Physical work involved on each item (in appropriate physical measure)	Physical work up to end of last year (as percentage or physical measure)	Physical target for the current year	Physical progress during the quarter	Physical work up to the end of the quarter as percentage of annual physical targets
....	....	....	....	....	....	....
....	....	....	....	....	....	....
....	....	....	....	....	....	....

5. Bottlenecks, hindering progress (Enter code number(s) from list below):
   (1) Delay in release of funds
   (2) Release of funds inadequate
   (3) Non-availability of foreign exchange components
   (4) Foreign experts not assigned
   (5) Administrative difficulties due to shortage/transfer of personnel
   (6) Workers' strikes
   (7) Non-selection of site
   (8) Delay in acquisition of land
   (9) Delay in preparation of detailed plans and estimates
   (10) Delay in designing of architects (*sic*)
   (11) Tenders for works not received
   (12) Approval of work tenders delayed
   (13) Delay by contractor in starting work
   (14) Contractor abandoned the work
   (15) Machinery not available locally
   (16) Delay in procurement of machinery
   (17) Difficulties in procurement of materials
   (18) Delay due to floods/rains
   (19) Other bottlenecks:
       Inter-departmental delays
       Delay in water connection
       Delay in electric connection
       Delay in gas connection
       Lack of popular participation at community level
       Lack of interest and initiative by community leaders
       Lack of locally mobilised funds
       Other delays

6. **Administrative status:**
   (a) Has there been any change in the scope of design of project? Yes/No
   (b) Is the project under revision?   Yes/No
   (c) Indicate whether there were any changes in senior management position.

   Number of personnel changed.

Project Director,
Other Director level,
Departmental Chief's level.

Fig. 1.2. Example of a quarterly progress report (traditional system).

ploughs, etc., are worked out by relatively sophisticated methods; costs are calculated by familiar budgeting methods; and yet, times for the completion of the project and its components are merely guessed. In this approach there is no equivalent of cost budgeting that can be applied to estimating durations of activities. Errors in estimating the times taken for various steps in the execution of the project can undermine the whole implementation process, for the reasons discussed in the previous section. Typically, time estimates are optimistic not only in the lengths of time allowed for the completion of particular activities, but also in failing to allow for the way in which one uncompleted activity can delay another. Although there are other reasons for this optimism, an important factor is that project promoters are optimistic people: they start out with the assumption that things can be improved (which is not such a bad thing).

(b) *Instruction*: The major problem is that the lack of any systematic method of converting the targets into instructions means that, on occasion, important instructions are not given in a timely manner to the right people. In addition, because there is no methodical checking of what has to be done, by whom, and when, the very existence of certain activities may be overlooked; and even if they are not overlooked, they may not be initiated. This probably applies more often to administrative activities than technical ones. For example, in one case, field staff had to pay for construction materials from accounts for which they were individually responsible. These accounts had to be regularly topped up with funds; however, the normal practice, built into the financial regulations—and into the perceptions of armies of clerks as to what was right and proper—was to top up only when the Treasurer's office issued a certificate that the account was exhausted. This inevitably resulted in delays, so a decision was taken to alter this procedure, at least for this one project. However, the detail of the alteration process was not analysed in sufficient detail, and the need for a critical authorisation was overlooked. As a result the project was regularly brought to a halt because the field staff had temporarily run out of funds.

(c) *Feedback*: The method of providing feedback suffers from three main disadvantages:

(i) A lot of the emphasis—all the emphasis in the monthly reports— is on financial measures. While these are obviously extremely important, they can be very misleading: almost everyone will be aware that 'progress', defined in these terms, can be inflated by getting materials delivered to a site, long before they can be used.

In addition, financial measurements alone will seldom yield reliable forecasts of future progress (see Chapter 8), because they are a consequence of physical activities, not independent events.

(ii) The physical progress summary tends to be a post-mortem, since it is recorded quarterly. Even when physical progress is to be recorded every month, deviations from schedule are often only picked up after they have become serious.

(iii) It gives no guidance to the project manager on the relative importance of project activities, at any one stage: except on the simplest projects, the manager cannot be concentrating on all the activities, all the time. His problem is that he has no way of picking out the steps that deserve special day-to-day attention; this is because no distinction is made between progress on key activities which can delay the whole project, and on non-urgent jobs which can suffer a certain amount of delay without serious effect.

A fourth disadvantage, closely related to the last point, is that this recording system provides no logical basis for summarising the physical state of progress. Without this, the form tends to become an undigested, unclassified list of everything that has happened since the last reporting date. This is no help to the project manager and a positive hindrance to senior officials in the project's sponsoring organisation. These men can expect a flow of detail that they will never be able to comprehend, on this system. What they need is information on key events of technical interest, and on key events which determine whether the project is remaining, or will remain, on' schedule. As we saw in an earlier section, the ability to provide this information is an important characteristic of a good project management system.

The system described above certainly lacks the last characteristic: the physical summary (so many metres of canal dug, so many acres of land levelled, so many doors put on field assistant's houses, so many filing cabinets purchased, etc.), contains so much detail that the overall picture is concealed. Even the head of, say, an agricultural department could find it difficult to dig out the implications for future progress of a project from such a report. An official further removed from the project could find it totally unintelligible. In the end, because it can be expressed in the fewest figures, the physical progress summary becomes replaced by the financial progress summary, and we are back where we were.

There is one positive feature of the system—the space for a 'bottle-

neck code' in the form shown in Fig. 1.2: however, in practice, because
it is so subjective, this tends to be used as a means of re-assigning blame,
rather than identifying problems.

Sadly, projects which are controlled in this way do display most of the
problems that the foregoing analysis suggests they would: key activities
are not initiated in time (or even not at all); times for whole groups of
activities prove to have been underestimated; the delays push costs up
and staff morale down; and the promoters of the project are repeatedly
surprised by the way in which forecast improvements in the rate of
progress, and of the effectiveness of remedial measures, fail to materialise.
Clearly, a better method is needed.

1.4. CRITICAL PATHS—INTRODUCTION TO AN IMPROVED METHOD

All the problems mentioned at the end of the previous section result from
the failure to plan and control the time needs of projects as carefully as
their money needs are budgeted and controlled. Unlike funding plans,
time schedules are generally neither feasible nor optimal; the instructions
for implementing items are neither clear nor explicit; and the feedback is
neither relevant nor adequately digested. The problems are not primarily
the result of the poor levels of productivity of staff in some under-
developed countries: in a culture in which physical work, decisions, and
payments are all made slowly (and possibly with good reason in terms of
poor health and small incentives) there is no chance of having a rapidly
executed project. There is only the choice between slow but organised
implementation, and chaos. The technique that will be outlined here
should do for time what budgets and financial control do for money, no
more: it will take realistic (usually existing) working methods and rates
to generate a good, feasible plan, and provide a means of implementing it.

Basically, the method consists of the following steps:

(a) assembly of a complete list of all the activities needed to complete
the project, together with the relationships between them (that is,
which job must follow which), and the lengths of time required to
complete them;
(b) calculation of the time needed to complete the project and the
dates on which each job should be started and finished;
(c) setting up of the instruction and feedback links between the
project manager and the executives.

1.5. AN EXAMPLE OF THE IMPROVED METHOD:
PLANNING THE SCHEDULE

To make the method clear, a simple example has been created, based on a real project, whose object was to set up an organisation to improve watercourses (i.e. the final channels in the irrigation system from which water is taken on to fields) and thus save water. The methods being used to attain this end were re-levelling and re-aligning each watercourse, lining part of it with concrete, and building into it proper culverts and pre-cast outlets. (The farmers had formerly simply breached the banks of the watercourse to get water.) To get the work going, it was necessary to acquire premises, purchase equipment, and train staff; Government was prepared to give grants to aid the work, but only through a system of cooperatives, which also had to be established. The example deals with planning the timing of this group of activities.

Table 1.1 shows the basic data (for the moment, we will ignore the whole complex business of obtaining this data and ensuring its accuracy, which will be discussed in Chapter 3).

It is not too difficult to plot these jobs to a time scale, as has been done in Fig. 1.3: this provides a graphical solution to the problem of calculating the project duration, and the starting dates for the activities. It also

Table 1.1. Basic Data for Simple CP Problem

Activity number	Activity description	Duration (weeks)	Activities which must precede it
1	Getting approval of project	12	—
2	Acquiring offices and training premises	6	1
3	Selecting and hiring the additional staff to be trained	6	1
4	Purchasing equipment for training and field work	4	1
5	Selecting staff to do the training and arranging their secondment	8	1
6	Selecting watercourses to be improved, and setting up cooperatives	10	1
7	Completing training in office	4	1–5
8	Completing field training	4	1–5, 7
9	Purchasing construction materials	4	1
10	Starting construction	—	1–9

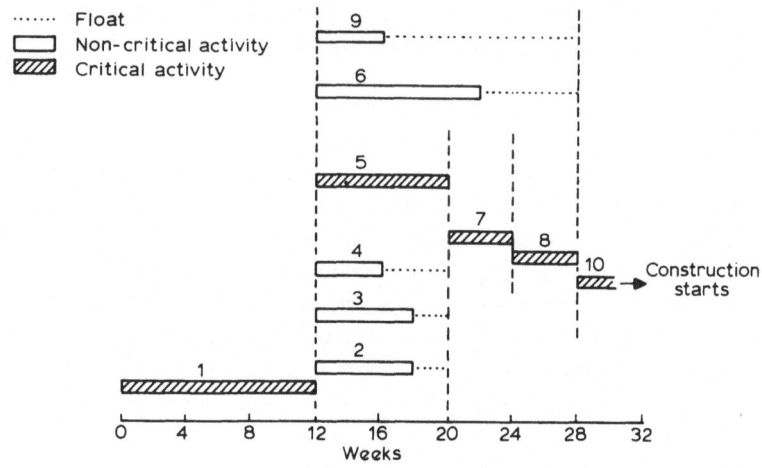

Fig. 1.3. Graphical solution to simple CP problem. See Table 1.1.

introduces the main concepts which will be elaborated on later. This figure was constructed by drawing a horizontal time scale (weeks) along the bottom of the figure, drawing the first activity (getting approval of the project) horizontally, and setting up a dotted vertical line at the end of it. All the activities that can only start after the first one is completed must start from this dotted line (which represents the start of the 13th week of the project). In this way, the durations of activities 2–6 and 9 can easily be plotted to scale. Activity 7 (the first of the two phases of training) is only slightly more difficult to plot: of the activities which must precede it, the latest to finish is activity 5 (selecting and seconding trainers); so activity 7 can start immediately after that, but not earlier. That is, it can start at the beginning of the 20th week. A second vertical dotted line has been drawn at that point: it determines the latest finishing date for activities 2–5, and shows clearly that there is more time available for activities 2, 3 and 4 than is required. (This spare time is called float.) Activity 8 can proceed after the latest finishing of activities 1–7, which is indicated by a further dotted line, at the end of activity 7. Finally, activity 10 can only start after the latest of all the others, which is activity 8, and the minimum time in which the project can be completed is 28 weeks.

Critical activities: Activities fall into two groups: those which have float associated with them, and those which do not. The latter are known as critical activities; delay in any of them will automatically lengthen the project. The continuous line that the critical activities form, from the start to the finish of the project, is known as the critical path (CP), and the methods of project control that depend on this sort of analysis are referred to as critical path methods (CPM). The length of the CP is the minimum time in which the project can be completed.

Non-critical activities: The non-critical activities, easily recognised because they have float, can suffer a certain amount of delay without holding up the whole undertaking, and a skilful project manager will exploit this. (If any one non-critical activity is delayed by more than the amount of float available, it will become critical, of course.)

Scope for applying CPM: Clearly we have the core of a method that meets the requirements set out in Section 1.2, and avoids the problems described in Section 1.3:

(a) Because of the systematic listing of activities, it is much less likely that any will be overlooked (activities would in practice be broken down into more sub-units than was done in the example).

(b) Provided that the time estimates for each activity are reasonably accurate, the overall schedule should be feasible, and the project manager will be aiming at a target that it is possible to hit.

(c) The instruction link in the control system can use the results from the diagram as a basis for issuing clear and specific instructions on when jobs should be started.

(d) The division into critical and non-critical activities enables the project manager's efforts to be concentrated on those activities which are most likely to delay the project; it also enables him to summarise data in a way that is useful to higher authorities (e.g. a government ministry for which an irrigation project is being implemented). Given a critical path analysis of a project, and up-to-date progress data, it is relatively easy to predict the effect of delay in any one activity on the whole project. Physical and financial summaries make more sense if they are accompanied by a soundly-based forecast of the future pattern of expenditure and physical progress.

One obvious deficiency of the method, as introduced above, is that the graphical approach used is inconvenient with very complex projects, if only because the vertical dotted lines (see Fig. 1.3), which show the interdependence of jobs, become difficult to interpret (because they may coincide with, or cut through, the horizontal bars representing unrelated activities). A more general method, arithmetic rather than graphical, will be introduced in the next chapter, and that will be used throughout the rest of this book. However, at least one major industrial concern (Monsanto) has used graphical CPM methods, and Lowe[1] gives a good description of these. (Some parts of the subsequent workings can most easily be done by re-converting the results to graphical form.)

Most of the Chapters 2–5 and 8 will be concerned with the detail required to turn the basic idea of CPM into a practical tool for managing and monitoring projects. The next chapter moves straight into the meat of the method. This is only partly the result of the writer's uncharitable assumption that you, the reader, will lose interest if not shown the actual works at an early stage: the definitions of some of the items of data will be easier to explain after showing how they are used.

1.6. CPM AND AFTER

In some fields in developed countries, there has recently been a move away from the use of CPM. It is significant that this is most evident in the building construction industry: with enough repetitions of very similar jobs, the transition from project to production is made. This means that, either as a result of previous CPM planning, or as a result of the evolution of effective working sequences under the pressure of past mistakes, the current working methods take account of the capacity for different project activities to interfere with each other. Similarly, plans and expectations embody reasonable expectations of work rates, and of the effects of the interactions between different jobs within a project, when this transition has been made. On development projects, because of their variety and the relative inexperience of many of the people involved, that transition is far in the future, if it can ever be expected.

A further factor is that many of the expectations of CPM have been unrealistic: particularly the expectation that the initial plan can be expected to hold good throughout project implementation—this is dealt with in detail in Chapter 8. Also, disappointment has been generated by the introductions of sophisticated improvements in the methodology (see Chapter 6) which impose unrealistic requirements on the quality of data that has to be gathered, and produce unrealistic answers as a result.

Most users of the simpler, more realistic tools in the CPM kit rapidly find them becoming part of their normal way of thinking: in deciding what to do next, most people and organisations have unconscious rules that they obey, such as 'do the easiest bit first', 'do the hardest bit first', 'always tidy up as you go', and so on. Someone who thinks in CPM terms will adopt the rule 'look for the thing that's holding the whole job back, and do that first'; he will also, as a matter of routine, have ensured he starts off with a complete list of all the tasks to be done, which is one of the most immediate advantages of the CPM approach.

REFERENCE

1. Lowe, C. W. (1969). *Critical Path Analysis by Bar-Chart*, Business Books, London.

CHAPTER 2

Solving the Critical Path Problem

2.1. THE EXAMPLE

2.1.1. Introduction

A slightly more complex example is introduced now, to provide a framework for developing the procedures needed to solve the twin problems of how long the project should take to complete, and what are the earliest and latest dates on which each activity should start and finish.

The example contains fewer activities than a real project would; the criteria for the amount of detail—in terms of the extent to which activities need to be broken down—are set out in the next chapter, when the reasons behind them will be easier to understand. However, all that we are interested in now is demonstrating the method, with a minimum of arithmetic complications.

It is worthwhile, at this stage, purely in the interests of giving you an indication of the relative scale of the problems, to compare the size of the example with some real projects. A purely agricultural project, such as setting up a new extension service, might consist of 100–150 activities; an agro–industrial project, with the construction of a processing plant included in it, could well consist of over 300 activities; whereas, our example will use under 50 activities.

Even so, larger projects, from the point of view of project control, only represent a greater mass of calculations to be handled: they do not introduce any new problems.

2.1.2. General Description of Example

The project chosen (a modified version of a real-life one from the writer's experience) is the setting up of a small dairy plant by a Department of

Agriculture; the plant was to be supplied with milk, partly from a government farm, and partly from farmers in the surrounding countryside. The development of the latter source of supply was seen as a longer-term objective, and not included in the 'project' as defined below. The following notes outline more clearly the nature of the undertaking:

(a) *Plant*: A small roller-drying plant was required; a contract for its construction was to be put out to tender overseas, as there were no manufacturers of suitable equipment in the country. A condition for pre-qualification was that the tenderers should have adequate experience of manufacturing and installing such equipment under similar conditions; enquiries elsewhere had established the lengths of time typically required for the manufacture and installation of comparable plants.

(b) *Buildings*: A relatively simple shell was required to house the plant, and the tender for its construction was to be put out locally.

(c) *Site*: The plant and farm were to be established on an existing government farm; however, establishment of new areas of forage, and fencing, improvements to the water supply, and internal roads were all required. The last three items were to be done by the Department's own labour force, and this is reflected in reduced mobilisation times for these items. Housing and milking facilities were already in existence.

(d) *Livestock*: These were to be imported, but not before the additional areas of forage had been established.

(e) *Farm equipment*: This was to be purchased locally, by the Department's usual tendering procedure.

For a more detailed description of a project, the form of visual representation referred to as a network is almost always used; its importance is so great that it deserves a section to itself.

2.2. THE NETWORK

2.2.1. Its Importance

A network is a diagram of a project, showing the inter-relationships of the activities which make it up. The name comes from the fancied resemblance to a fish net, which may just be apparent in Fig. 2.1. An alternative name is arrow diagram, for obvious reasons.

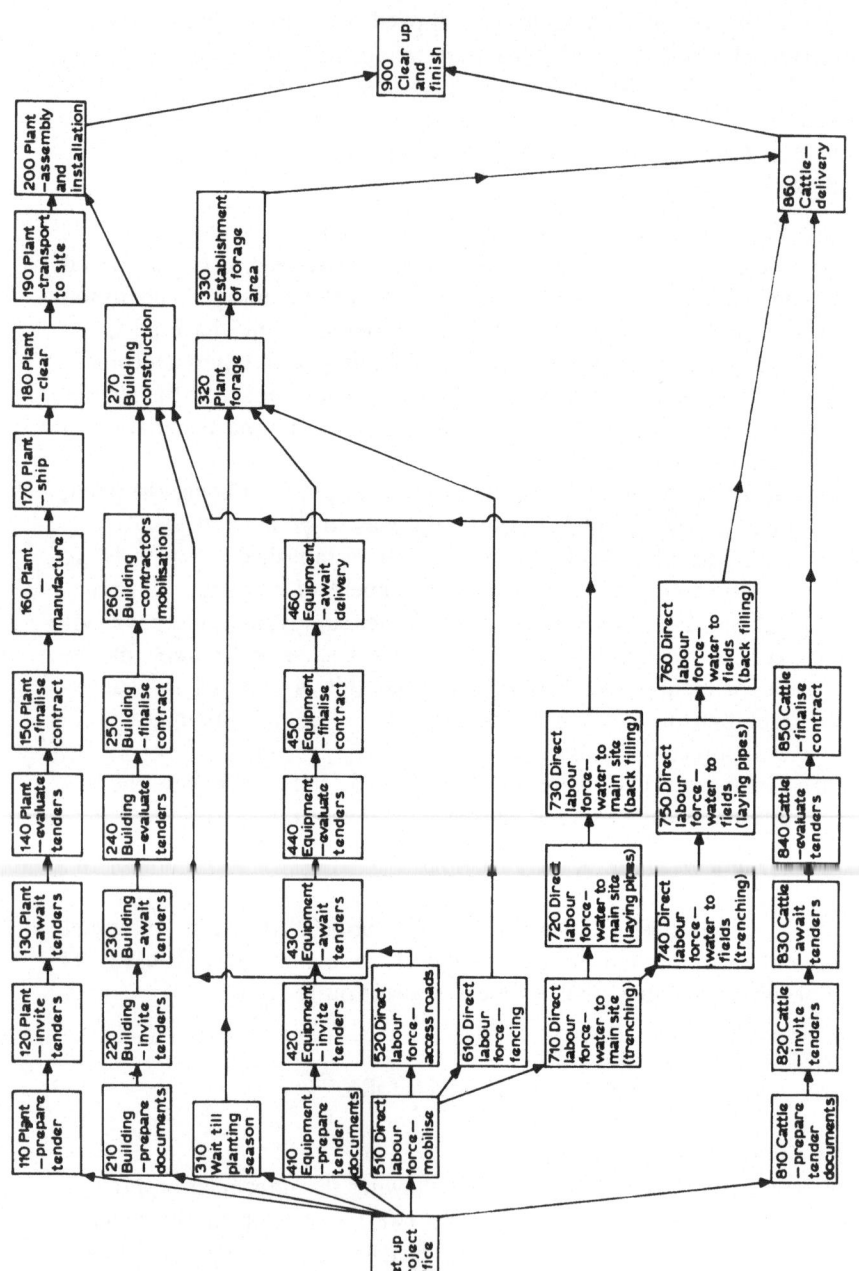

Fig. 2.1.   Dairy project network.

There are various methods of drawing networks, the most significant division being between those in which the activities are represented by the shafts of the arrows, and those in which they are written at the ends of the arrows. For reasons that will become apparent later, the latter method of representation is usually easier for people with no CPM background to read and interpret correctly. It is therefore also easier for them to contribute to. This is even more important, for the following reason. Usually, the person carrying out the CP analysis will not be technically qualified in all aspects of the project; he will have to extract the information he requires from those who are. The more the latter can contribute their special knowledge by participating in the drafting of the network, the better, since it is the starting point for all the subsequent operations: the lists of activities, durations, etc., are compiled from the network—not the other way round. If the network is not right, then the foundation of the whole exercise is weakened.

Figure 2.1 shows the network for the example project introduced in Section 2.1. Before going on to demonstrate the method in detail, it is worth looking at it as an example of networks in general, to see how they are put together.

2.2.2. General Principles of Constructing Networks

Networks such as this—where the activities are shown at the ends of the arrows–are described as 'activity-on-node' networks because the activities are written at the 'knots' in the 'net' (node comes from the Greek for knot).

These knots are the points where the arrows representing the linkages between different activities end, converge, or diverge. All that 'linkage' implies is that there is a relationship between the activities at each end of the arrow; the only type of linkage that can be directly shown is that one or more activities must follow one or more others. The preceding activity is always at the 'blunt' end of the arrow, the succeeding one, at the 'sharp' end. Examples in Fig. 2.1 include:

(a) *Simple sequence*: This occurs when one activity must follow another, e.g. activity 120 (advertising for tenders for the plant) must be preceded by activity 110 (preparing the tender documents).

(b) *Convergence*: This occurs when several activities must be completed before one can start, e.g. activity 860 (delivery of cattle) must follow all of the following: completing water supply (760), establishing forage area (330), and completing negotiations for the

supply of cattle (850). Therefore, an arrow points from each of
these to activity 860 (supply of cattle).

(c) *Divergence*: This occurs when several activities are all dependent
on the completion of one other, e.g. activities 110, 210, 410 and 810
(preparing tender documents for plant, construction of buildings,
locally purchased equipment, and cattle), and activity 510 (mobilis-
ing direct labour force) can only start after the project office has
been set up.

It will also be useful to define a segment of the network as any simple,
unbranched sequence starting and ending at points that are either a
convergence or a divergence (activities 410–460 form a segment, and a
segment may consist of only one activity), and a path as any continuous
sequence of activities.

As a matter of practical tactics, it is always sound to ensure that the
flow of the network is in the same direction throughout, i.e. all the
arrows point either from left to right, or down the page. It is, of course,
perfectly possible to represent a project unambiguously without this
restriction. However, without it, it is easier for the reader to misinterpret
what he sees: he may mistakenly assume that an activity to the right of
(or below) another to which it is linked must follow it, if he does not
check which way the arrow points.

This restriction does mean a little care is necessary in drafting the
network, but this is eased by the fact that the length of the arrows is
irrelevant. Unlike the presentation in Fig. 1.3, the arrows are not drawn
to a time scale; Fig. 2.2 illustrates the differences between the two
methods of representation.

At first sight, it seems that we are dealing with only a very limited
selection of all the possible relationships between activities: that one (or
more) can only start after one (or more) others. The next section shows
how more complex relationships can be handled.

2.2.3. More Detail on Constructing Networks
There are a variety of conceivable relationships, other than the 'must
follow' one:

(a) One activity must start after another one has progressed a little
way. An example of this occurs in Fig. 2.1: laying pipe (activities 720 and
750) can only start after trenching (activities 710 and 740) is started and
has progressed some way. This is handled by breaking down the
trenching into segments of reasonable length, from the operational point

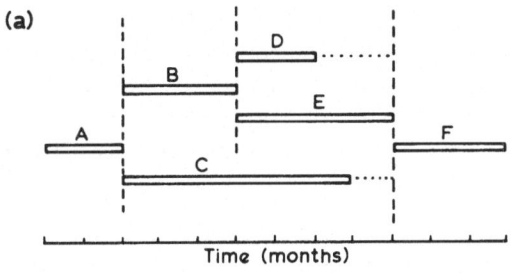

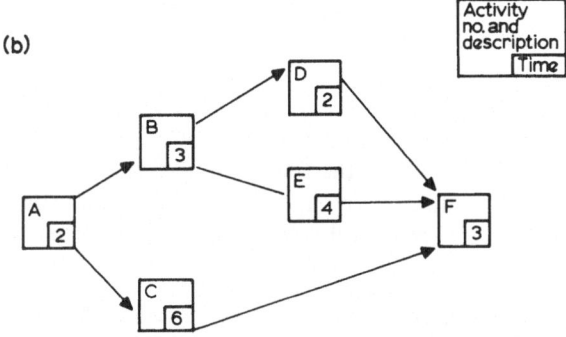

Fig. 2.2. Two versions of a network: (a) time scale; (b) activity-on-node.

of view, and linking them in the manner shown. Here, the segments are 'borehole to main site' and 'main site to fields'. The representation in Fig. 2.1 assumes that the sections must be dug in sequence; if this was not so, activities 710 and 740 would both branch from activity 510, and whether they would be done simultaneously or not would be determined by the number of diggers available (see Chapter 4, where the problem of resource constraints is dealt with).

(b) One activity can only start after a certain date, e.g. activity 320 (planting the forage area) can only start after the start of the planting season. An artificial waiting activity (310) is inserted before the planting activity with its length calculated to ensure that the planting cannot start too early; this implies that the calculations (see below) must be made relative to a fixed starting date on the calendar, of course.

(c) An activity has to be completed less than a certain length of time before another starts. This occurs where the preceding activity of the pair produces something perishable, e.g. it might be 'deliver cement'. Also tenders are perishable: they are only valid for a certain period; so, activity 850 (completing negotiations of tender for livestock) should not finish more than three months before activity 860 (delivery of cattle) starts. This limitation cannot be incorporated directly in the network, but is easily dealt with later in the calculations (see Chapter 4); such activities should be distinguished in the network.

(d) An activity has to be completed by a certain date: methods have been proposed for incorporating so-called 'scheduled dates' into networks. This may lead to problems: the network may represent an insoluble problem, or there may be odd isolated sequences of critical activities, separated from the main critical path. Methods have been proposed for dealing with these eventualities, but it seems to the writer that there is a basic inconsistency in attempting to impose such scheduled dates at all. If you are reasonably sure of the order in which activities are to be done and the times required for each, CPM will tell you whether the schedule will be met or not, and if not, provide you with the means to decide what modifications are necessary to meet it (Chapter 5).

(e) One or more alternative paths may be followed, according to the outcome of some event; for example, if one activity is getting confirmation of the appointment of a senior member of staff from a Public Services Commission, there are two possible pathways leading from it. If he is accepted, he is mobilised; if not, selection has to be restarted. There is a method of handling this, using what are called decision networks, but it is seldom applied in practice; Chapter 6 gives a reference to a more advanced source of information on this topic, and includes a brief introduction, with a worked example.

(f) Two (or more) activities can be carried out in either order, but not simultaneously; this is always because they compete for some resource, even if it is only working space or access, and this situation can be handled by the resource scheduling methods described in Chapter 4.

The last three types of relationship are included for completeness—they do not appear in our example.

Now, with a reasonably detailed description of the example, and an understanding of the way it is represented in the network, we can return to the main theme: a non-graphical method of solving the twin problems of finding the critical path, and of determining the best starting and finishing dates for all activities.

2.3. THE MAIN OBJECTIVE: FINDING THE CRITICAL PATH

It is worth recalling why we are attempting to do this: the CP is the chain of activities that determines the duration of the project, and the activities that form its links differ significantly from the others. Non-critical activities can be shifted about, to ease the burden on key resources; critical ones can be given special attention by the project manager, and are of special interest when the question of accelerating the project arises, since these are the ones whose duration has to be shortened.

2.4. TABULATION OF THE DATA

2.4.1. Format

The information shown on the network—which is the primary source of the data, for reasons that will become apparent later—is written out in tabular form. This has been done for our dairy scheme in Table 2.1. The columns are:

(A)   *Activity number*: these are for cross-reference purposes in the table, and can be allocated arbitrarily. However, it makes the significance of the numbers easier to remember if all the numbers of activities relating to a particular technical function start with the same digit. Also, it is safer to number the activities in steps of 10 rather than in steps of 1, since after-thoughts are then more easily inserted.

(B)   *Brief description*: only included to save the user from referring back to the network to find out to what activities the numbers refer.

(C)   *'Follows' column*: contains the numbers of those activities that must be completed before the current one starts.

(D)   *'Precedes' column*: contains the numbers of those activities that can only start after the current one is completed (obviously, this conveys no more information than (A) and (C) together, but it makes the calculation easier).

(E)   *Duration*: the length of time that the activity takes to complete.

(F–H) *Early start and finish*; *late start and finish*; *and float* (spare time available). These last six columns will contain the results of the calculations.

Table 2.1. Dairy Project Data

Activity number (A)	Description (B)	Follows (C)	Precedes (D)	Duration (E)	Early Start (F1)	Early Finish (F2)	Late Start (G1)	Late Finish (G2)	Total Float (H1)	Free Float (H2)
1	Set up project office	—	110, 210, 310, 410, 510, 810	1						
110	Plant—prepare tender documents	1	120	2						
120	Plant—invite tenders	110	130	1						
130	Plant—await tenders	120	140	4						
140	Plant—evaluate tenders	130	150	1						
150	Plant—complete negotiations and signing contract	140	160	3						
160	Plant—manufacture	150	170	20						
170	Plant—ship	160	180	6						
180	Plant—clear	170	190	1						
190	Plant—transport to site	180	200	2						
200	Plant—assembly and installation	190, 270	900	22						
210	Building—prepare tender documents	1	220	2						
220	Building—invite tenders	210	230	1						
230	Building—await tenders	220	240	3						
240	Building—evaluate tenders	230	250	1						
250	Building—negotiations and signing contract	240	260	2						
260	Building—contractors mobilisation	250	270	4						
270	Building—construction	260, 730, 520	200	30						
310	Wait till planting season	1	320	31						
320	Plant forage area	610, 310, 460	330	4						
330	Forage area becoming established	320	860	15						
410	Locally purchased equipment—draft specifications	1	420	1						
420	Locally purchased equipment—advertise	410	430	1						

430	Locally purchased equipment—await tenders	420	440	3
440	Locally purchased equipment—evaluate tenders	430	450	1
450	Locally purchased equipment—negotiations and signing of contract	440	460	1
460	Locally purchased equipment—await delivery	450	320	5
510	Direct labour force—mobilise	1	520, 610, 710	4
520	Direct labour force—build road	510	270	9
610	Direct labour force—fencing	510	320	12
710	Direct labour force—piped water supply to main site, trenching	510	720, 740	3
720	Direct labour force—piped water supply to main site, laying	710	730	2
730	Direct labour force—piped water supply to main site, back filling	720	270	1
740	Direct labour force—piped water supply to fields, trenching	710	750	3
750	Direct labour force—piped water supply to fields, laying	740	760	2
760	Direct labour force—piped water supply to fields, back filling	750	860	1
810	Cattle—prepare tender documents	1	820	2
820	Cattle—invite tenders	810	830	1
830	Cattle—await tenders	820	840	4
840	Cattle—evaluate tenders	830	850	2
850	Cattle—negotiations and signing contract	840	860	2
860	Cattle—delivery	850, 760, 330	900	4
900	Job finishes	200, 860	—	1

2.4.2. Checking
There are a couple of useful checks that can be applied to the table at this stage:

(a) provided that there is a single starting and a single finishing activity, the number of entries in the 'follows' column must equal the number of entries in the 'precedes' column;

(b) the number of entries in the 'follows' column should equal the number of arrows in the network.

2.5. THE FORWARD PASS—CALCULATING EARLY START AND EARLY FINISH DATES

'Early start' and 'early finish' are shorthand phrases for the earliest possible start and finish, respectively, of each activity; unless non-critical activities have to be delayed because of resource limitations, these are the dates on which each activity should start and finish—since any unnecessary delay increases the chances of something going wrong.

The principle of the calculation is simple: the early start date of any one activity is fixed by the durations of all the activities that must precede it. Looking at the network, there seem to be three distinct cases, which will now be considered below.

(a) *Divergences*: where arrows diverge, more than one activity is dependent on the activity under consideration. An example of this occurs right at the beginning of the network. Activity 1 (set up project office) starts at the beginning of week 1, takes one week, and therefore ends at the end of week 1. Anything that follows this activity cannot start before week 2, so week 2 is entered as the early start for activities 110, 210, 310, 410, 510 and 810. In each case, the early finish is one unit (week) less than the early start of the next activity, because we are using inclusive dates: a three-week activity starting in week 6 ends in week 8, having run through weeks 6, 7 and 8; and a one-week activity starting in week 1 ends in week 1, as we have seen.

(b) *Sequences*: where only one activity follows only one other. The rule here is obvious: if an activity takes N units of time, starting on date D, then its early finish is on $D + N - 1$, and the next activity's early start is $D + N$. (Remember we are working with inclusive dates.) These dates should be written in the appropriate spaces on the row for the activity or activities in the F1 and F2 columns. For example, in the sequence of activities 210–260, concerned with preparation to start building, the early start of activity 210 is week 2 and the activity takes two weeks; its early finish is therefore week $2+(2-1)=3$, and the next activity has its early start in week $2+2=4$. In this way, you should be able to calculate the early finishes for activity 260 as week 14, for activity 520 (access road) as week 14, and for activity 730 (completion of water supply to main site) as week 11.

(c) *Convergences*: where one activity is dependent on the completion of more than one preceding activity. An example is activity 270 (construction of building for plant), which can only start after activities 260, 520 and 730. If you have followed the procedure suggested above, you will have three entries pencilled in as possible early starts: weeks 14, 14 and 11. Clearly, if activity 270 has to wait until all the activities directly preceding it are completed, it cannot start before the latest of the early starts pencilled in, i.e. week 14.

Incidentally, there is a useful check which can be made, if the format shown in Table 2.1 is used: the number of possible early start dates pencilled in at an activity at a convergence must equal the number of activities in the 'follows' column. If this is not so, there must be another earlier segment of the network for which the calculations have to be done, before the early start at the convergence can be calculated.

Table 2.2 summarises the rules for the forward pass calculation:

Table 2.2. Forward Pass Calculations

Situation	Current activity			Next activity		
	Duration	Early start	Early finish	Duration	Early start	Early finish
Divergence from current activity	N	D	$D+(N-1)$	M	$D+N=D1$	$D1+(M-1)$
Straight sequence	N	D	$D+(N-1)$	M	$D+N=D1$	$D1+(M-1)$
Convergence on to next activity	N'	D'	$D'+(N'-1)$	M	D1 is the latest of:	$D1+(M-1)$
	N''	D''	$D''+(N''-1)$		$\left\{ \begin{array}{c} D'+N' \\ D''+N'' \\ D'''+N''' \end{array} \right\}$	
	N'''	D'''	$D'''+(N'''-1)$			

Table 2.3. Dairy Project Results of Forward and Backward Passes

Activity number (A)	Description (B)	Follows (C)	Precedes (D)	Duration (E)	Early Start (F1)	Early Finish (F2)	Late Start (G1)	Late Finish (G2)	Total Float (H1)	Free Float (H2)
1	Set up project office	—	110, 210, 310, 410, 510, 810	1	1	1	1	1		
110	Plant—prepare tender documents	1	120	2	2	3	5	6		
120	Plant—invite tenders	110	130	1	4	4	7	7		
130	Plant—await tenders	120	140	4	5	8	8	11		
140	Plant—evaluate tenders	130	150	1	9	9	12	12		
150	Plant—complete negotiations and signing contract	140	160	3	10	12	13	15		
160	Plant—manufacture	150	170	20	13	32	16	35		
170	Plant—ship	160	180	6	33	38	36	41		
180	Plant—clear	170	190	1	39	39	42	42		
190	Plant—transport to site	180	200	2	40	41	43	44		
200	Plant—assembly and installation	190, 270	900	22	45	66	45	66		
210	Building—prepare tender documents	1	220	2	2	3	2	3		
220	Building—invite tenders	210	230	1	4	4	4	4		
230	Building—await tenders	220	240	3	5	7	5	7		
240	Building—evaluate tenders	230	250	1	8	8	8	8		
250	Building—negotiations and signing contract	240	260	2	9	10	9	10		
260	Building—contractors mobilisation	250	270	4	11	14	11	14		
270	Building—construction	260, 730, 520	200	30	15	44	15	44		
310	Wait till planting season	1	320	31	2	32	13	43		
320	Plant forage area	610, 310, 460	330	4	33	36	44	47		
330	Forage area becoming established	320	860	15	37	51	48	62		
410	Locally purchased equipment—draft specifications	1	420	1	2	2	32	32		
420	Locally purchased equipment—advertise	410	430	1	3	3	33	33		

Activity	Description	Preceding	Succeeding					
430	Locally purchased equipment—await tenders	420	440	3	4	6	34	36
440	Locally purchased equipment—evaluate tenders	430	450	1	7	7	37	37
450	Locally purchased equipment—negotiations and signing of contract	440	460	1	8	8	38	38
460	Locally purchased equipment—await delivery	450	320	5	9	13	39	43
510	Direct labour force—mobilise	1	520, 610, 710	4	2	5	2	5
520	Direct labour force—build access road	510	270	9	6	14	6	14
610	Direct labour force—fencing	510	320	12	6	17	32	43
710	Direct labour force—piped water supply to main site, trenching	510	720, 740	3	6	8	9	11
720	Direct labour force—piped water supply to main site, laying	710	730	2	9	10	12	13
730	Direct labour force—piped water supply to main site, back filling	720	270	1	11	11	14	14
740	Direct labour force—piped water supply to fields, trenching	710	750	3	9	11	57	59
750	Direct labour force—piped water supply to fields, laying	740	760	2	12	13	60	61
760	Direct labour force—piped water supply to fields, back filling	750	860	1	14	14	62	62
810	Cattle—prepare tender documents	1	820	2	2	3	52	53
820	Cattle—invite tenders	810	830	1	4	4	54	54
830	Cattle—await tenders	820	840	4	5	8	55	58
840	Cattle—evaluate tenders	830	850	2	9	10	59	60
850	Cattle—negotiations and signing contract	840	860	2	11	12	61	62
860	Cattle—delivery	850, 760, 330	900	4	52	55	63	66
900	Job finishes	200, 860	—	1	67	67	67	67

By repeatedly applying these rules you should be able to verify that columns F1 and F2 in Table 2.3 are correct. The early start for 'job finished' is week 67, and the project therefore will take (at least) 66 weeks to complete.

2.6. THE BACKWARD PASS—CALCULATING LATE START AND LATE FINISH DATES

'Late start' and 'late finish' are, respectively, shorthand phrases for the latest starting and finishing dates that will allow the project to be completed at the earliest possible date. The process starts from project completion (week 67 in this case), and works back; this is the reason why it is referred to as the backward pass. It is the mirror image of the forward pass, as consideration of the three cases will show:

(a) *Convergences*: (one activity dependent on more than one activity) as at the beginning of activity 200 (assembly and fabrication of the plant): activity 200 must finish at latest in week 66. As it takes 22 weeks, it must start by week $66 - (22 - 1) =$ week $(66 - 21) =$ week 45, since we are still using inclusive dates. All the activities which directly precede activity 200 must therefore finish in week 44, which is entered against each as its late finish: activity 190 (transport plant to site) and activity 270 (complete building) must be completed by week 44.

(b) *Straight sequences*: the rule here is that, if an activity has a late finish date of D and takes N weeks, then the preceding activity has its late finish on $D - N$, and itself has a late start of $(D - N) + 1$. For example, activity 190 (transport plant to site) has its late finish on week 44, and takes two weeks; it must therefore have its late start on week $(44-2) + 1 = 43$, and its preceding activity, 180 (clearance of plant through port) has a late finish at week $(44-2) = 42$. These dates are entered in the appropriate place in the table (columns G1 and G2).

(c) *Divergences*: (more than one activity dependent on one activity), as occurs at the end of activity 510 (mobilisation of direct labour force). As in the case of convergences in the forward pass, where divergences occur, a number of alternative late finishes will be pencilled in. If you have been following the procedure described above, you will have three possible dates pencilled in as possible late finishes for this activity: week 5 (from activity 520), week 31 (from activity 610), and week 8 (from activity 710).

By definition, the late finish of activity 510 must be early enough not to delay any of these three activities, so it is the earliest of the late finishes that is used. Table 2.4 pulls these rules together in summary form, and comparison with Table 2.2 will show the extent to which the two sets of rules are a neat reversal of each other. Even the check is reversed: there must now be as many late finishes pencilled in against each activity as there are entries in its 'precedes' column. You should be able to follow this procedure through, to complete all of columns F1 to G2 in Table 2.1, to verify Table 2.3.

Table 2.4. Backward Pass Calculations

Situation	Current activity			Preceding activity		
	Dura-tion	Late finish	Late start	Dura-tion	Late finish	Late start
Divergence	N'	D'	$D'-N'+1$	M	D1 is the earliest of $\left\{\begin{array}{l} D'-N' \\ D''-N'' \\ D'''-N''' \end{array}\right\}$	$D1-M+1$
	N''	D''	$D''-N''+1$			
	N'''	D'''	$D'''-N'''+1$			
Straight sequence	N	D	$D-N+1$	M	$D-N=D1$	$D1-M+1$
Convergence	N	D	$D-N+1$	M	$D-N=D1$	$D1-M+1$

2.7. FLOAT

Float is the amount of time that the start (and therefore the finish) of an activity can be delayed, without project completion time being lengthened. The types of float are described below.

2.7.1. Total Float
This is the amount by which any activity can be delayed, without affecting project completion time; by the definition of early and late start, it is the difference beween these two quantities; it is also equal to the difference between the early and late finish of the activity, too, and this provides a useful check on the arithmetic. You should be able to go through Table 2.5 and check the calculation of total float. You will notice there is an odd pattern in the floats: they are all of equal size for all activities on the same segment of the network. A little thought will

Table 2.5. Dairy Project Results, Including Float

Activity number (A)	Description (B)	Follows (C)	Precedes (D)	Duration (E)	Early Start (F1)	Early Finish (F2)	Late Start (G1)	Late Finish (G2)	Total Float (H1)	Free Float (H2)
1	Set up project office	—	110, 210, 310, 410, 510, 810	1	1	1	1	1	0	0
110	Plant—prepare tender documents	1	120	2	2	3	5	6	3	0
120	Plant—invite tenders	110	130	1	4	4	7	7	3	0
130	Plant—await tenders	120	140	4	5	8	8	11	3	0
140	Plant—evaluate tenders	130	150	1	9	9	12	12	3	0
150	Plant—complete negotiations and signing contract	140	160	3	10	12	13	15	3	0
160	Plant—manufacture	150	170	20	13	32	16	35	3	0
170	Plant—ship	160	180	6	33	38	36	41	3	0
180	Plant—clear	170	190	1	39	39	42	42	3	3
190	Plant—transport to site	180	200	2	40	41	43	44	3	3
200	Plant—assembly and installation	190, 270	900	22	45	66	45	66	0	0
210	Building—prepare tender documents	1	220	2	2	3	2	3	0	0
220	Building—invite tenders	210	230	1	4	4	4	4	0	0
230	Building—await tenders	220	240	3	5	7	5	7	0	0
240	Building—evaluate tenders	230	250	1	8	8	8	8	0	0
250	Building—negotiations and signing contract	240	260	2	9	10	9	10	0	0
260	Building—contractors mobilisation	250	270	4	11	14	11	14	0	0
270	Building—construction	260, 730, 520	200	30	15	44	15	44	0	0
310	Wait till planting season	1	320	31	2	32	13	43	11	0
320	Plant forage area	610, 310, 460	330	4	33	36	44	47	11	0
330	Forage area becoming established	320	860	15	37	51	48	62	11	0
410	Locally purchased equipment—draft specifications	1	420	1	2	2	32	32	30	0
420	Locally purchased equipment—advertise	410	430	1	3	3	33	33	30	0

430	Locally purchased equipment—await tenders	420	440	3	4	6	34	36	30	0
440	Locally purchased equipment—evaluate tenders	430	450	1	7	7	37	37	30	0
450	Locally purchased equipment—negotiation and signing of contract	440	460	1	8	8	38	38	30	0
460	Locally purchased equipment—await delivery	450	320	5	9	13	39	43	30	19
510	Direct labour force—mobilise	1	520, 610, 710	4	2	5	2	5	0	0
520	Direct labour force—build access road	510	270	9	6	14	6	14	0	0
610	Direct labour force—fencing	510	320	12	6	17	32	43	26	15
710	Direct labour force—piped water supply to main site, trenching	510	720, 740	3	6	8	9	11	3	0
720	Direct labour force—piped water supply to main site, laying	710	730	2	9	10	12	13	3	0
730	Direct labour force—piped water supply to main site, back filling	720	270	1	11	11	14	14	3	3
740	Direct labour force—piped water supply to fields, trenching	710	750	3	9	11	57	59	48	0
750	Direct labour force—piped water supply to fields, laying	740	760	2	12	13	60	61	48	0
760	Direct labour force—piped water supply to fields, back filling	750	860	1	14	14	62	62	48	37
810	Cattle—prepare tender documents	1	820	2	2	3	52	53	50	0
820	Cattle—invite tenders	810	830	1	4	4	54	54	50	0
830	Cattle—await tenders	820	840	4	5	8	55	58	50	0
840	Cattle—evaluate tenders	830	850	2	9	10	59	60	50	0
850	Cattle—negotiations and signing contract	840	860	2	11	12	61	62	50	39
860	Cattle—delivery	850, 760, 330	900	4	52	55	63	66	11	11
900	Job finishes	200, 860	—	1	67	67	67	67	0	0

show that this must be so: on such a path, each successive late start will be the same as that of the succeeding activity, less the succeeding activity's duration; and each early start will be the same as the preceding early start, plus the preceding activity's duration. Only where there are convergences can there be a choice—and therefore possibly a change—of early start (Section 2.5); and only where there are divergences can there by a choice of late finish (Section 2.6).

Total float therefore is a property of the whole network, and the amounts shown do not mean that *every* activity with float can be delayed by that much without affecting project completion time. If, for example, the float of 30 weeks on activity 410 (draft specifications for locally produced equipment) is used by delaying its start to week 32, there will be *no* float available for activities 420, 430, 440, 450 and 460. Despite this, total float is of far more use in resource scheduling (Chapter 4) than any other. (Resource scheduling is the process of re-arranging start and finish dates to fit resource limitations.)

2.7.2. Free Float

This is the float available when all jobs start as early as possible: i.e. instead of taking 'late finish' minus 'early finish', we take 'early start of succeeding activity' minus 'early finish of the activity'. This reflects better the amounts of 'spare' time available in a tightly-run project, but the free floats are *still* a property of the network: the actual float of any activity will be affected by the amount of float used up by the preceding activities.

Because of the way the calculations are carried out, only the last activity in each segment of the network will have free float, since in a segment, on a straight sequence, each activity's early start follows immediately on the early finish of its immediate predecessor. However, free float does have a use: as a warning during resource scheduling, where the free float corresponds to the maximum period by which any activity can be delayed without requiring other activities to be re-scheduled as a consequence.

You should verify the calculations of free float in Table 2.5, and check them against Fig. 2.3.

2.7.3. Interfering Float

If total float is the amount by which an activity's start can be delayed without affecting project completion, and free float is the extent to which the activity's start can be delayed without necessitating any re-scheduling

Solving the Critical Path Problem

39

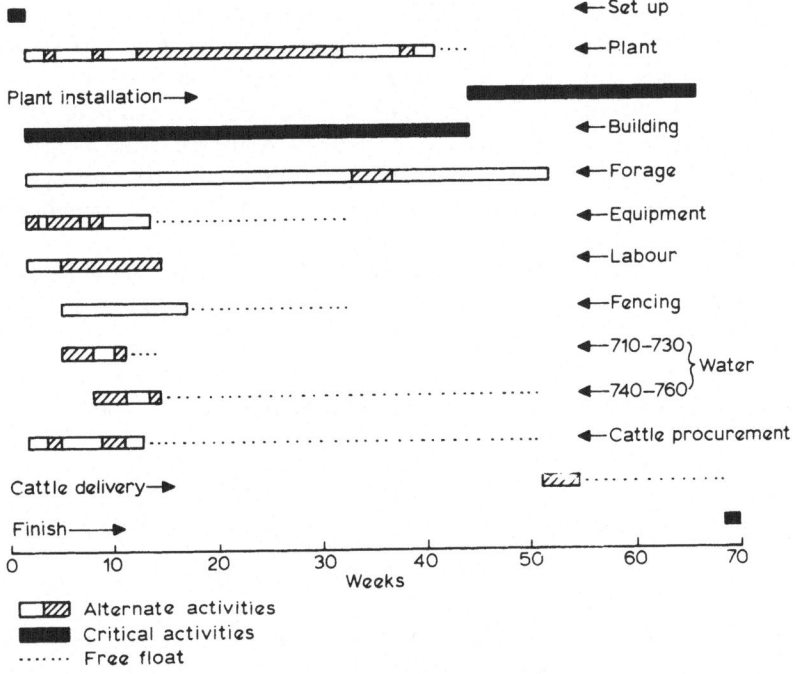

Fig. 2.3. Bar-chart of dairy project.

of other activities, the difference between the two must be the measure of the amount of 'downstream' disturbance caused by moving any activity to the extent of its total float. This is interfering float.

Since free float is zero along any segment of the network, at any point in a segment, all the float is interfering float; only at convergences and divergences will it differ from total float; and all the interfering float reappears 'downstream' as free float at some point. (Think what would happen when you get to the end of the network if this was not true.) Interfering float is not sufficiently useful to be worth tabulating.

2.7.4. Independent Float

In many ways, the most interesting property of independent float is a psychological one: most people with a slight acquaintance with CPM believe that float means the same thing as independent float. The danger of this will become apparent later.

An activity's independent float is the minimum amount of spare time available, over and above that necessary to complete it, regardless of the start dates of other activities in the network (provided they start earlier than their late start dates, of course). It is the difference between each activity's late finish and the early start of its successor(s), and really does represent potential idling time for the activity. And here is the danger in publicizing the amounts of float among project staff: they are apt to assume that *any* float is independent float, and adopt a correspondingly slack attitude to scheduled starting dates.

It is unusual for there to be independent float on many activities in a network, which limits its usefulness as a criterion for ranking activities for allocation of scarce resource scheduling.

When the independent float is less than zero, it means that it is not possible to delay the activity until its latest start *and* to start all the subsequent ones at their earliest starts.

2.8. THE CRITICAL PATH

There will always be one path, from start to finish through the network, that has no float; this is the critical path, and it determines job duration. You will now be able to pick it out in Table 2.3 as the sequence of jobs that have no total float. In this instance it branches, but it must always form a continuous chain from start to finish. If not, you have made a mistake!

It is important that at this stage, the network is 'read' carefully, to determine whether the result of the analysis makes sense:

(a) Is the critical path covering activities that past experience suggests are likely to hold up projects? If not, are the input data faulty, or are there special circumstances that make the past experience irrelevant or unreliable in this instance?

(b) Is the overall completion time reasonable, in view of comparable projects ('reasonable' does not necessarily mean 'within the specified limit')?

(c) If the critical path passes these tests, are there other near-critical segments that need special attention? ('Near critical' can usefully be taken to mean having a total float of less than 10% of its total duration—although, obviously, this depends on the reliability of the data, in the way described in Chapter 6.)

Once these questions have been answered, it is necessary to present the results of the analysis to the heads of the sections involved, for review, before proceeding to the next stage.

2.9. REVIEW OF THE INITIAL ANALYSIS

A good method of presentation is to convert the network to a bar-chart form: a time scale is drawn, by working days, and the activities are plotted on it, by earliest start dates. (This has been done in Fig. 2.3 for our dairy scheme example. The vertical spacing is used to distinguish particular groups of activities, e.g. by departments.) This can usefully be presented at a discussion, with the following agenda:

(a) The total period required to complete the project.

(b) The time at which activities are currently scheduled to take place: Do these present seasonal (e.g. no excavation in flood season) or other problems?

(c) The initial resource requirements: heads of specialities are asked to identify major potential clashes of resource requirements (e.g. peaks of labour demand) apparent from the bar-chart.

It is important that the meeting is tactfully reminded that the results presented are only the logical consequences of the data which its members have fed to the analyst. If this point is not established, he may be over-ridden by objections whose roots are in existing undertakings to complete the project by a certain date: there is a danger that arbitrary revisions to working rates, or the network, may be made, with the object of ensuring that estimated completion time matches the target. This may stifle immediate conflict with the project's sponsors, but it is obviously no long-term solution. This sort of trouble can best be avoided by introducing ideas for accelerating project completion using the techniques to be discussed in Chapter 5. These include changing the job specification, 'crashing' (selectively allocating extra resources to) certain activities, and changes in working practices.

Nevertheless, the heads of sections must also be made aware that the next phase—resource scheduling—may make the picture even blacker, if the initial analysis has produced an unexpectedly late completion date.

However, before we look at resource scheduling, we must take a step back to look at some alternative methods of drawing and specifying networks (Section 2.10), and, much more important, at the problems of actually getting the sort of data shown in Table 2.1; this will be the subject of the next chapter.

2.10. ALTERNATIVE METHODS OF NETWORK CONSTRUCTION

As has already been indicated, there are alternative methods of specifying and drawing networks. The most important alternative presentations are:

Activity-on-arrow networks
Precedence diagrams
Event networks
Gantt charts

These are dealt with in turn in the remainder of this section.

2.10.1. Activity-on-Arrow Networks

The basic difference from the method we have been using is obvious from the name: the identities and durations of the activities are written on the shafts of the arrows. The lengths of the arrows are arbitrary, and do not represent the durations of the activities. Figure 2.4 shows a small piece of network (the same one as used in Fig. 2.2) in both activity-on-node and activity-on-arrow formats.

In this scheme, the nodes are called events: these are end points—beginnings or completions—of activities. The events may be numbered, and the activities named by two numbers, the starting and finishing event numbers as in Fig. 2.4(c). If this is done, additional dummy activities may be needed to avoid confusion, e.g. between activities 5–3 and 2–3.

Activity-on-arrow networks are popular as the basis for computer programs, and many CPM-based project management programmes require that the input be specified in this way. However, the method has no particular advantage over the one chosen here, and it has one significant disadvantage. Look at the network shown in Fig. 2.5(a) in activity-on-node format, and consisting of six activities. The first activity, O, can be translated readily enough; and activities A and B must clearly start from its end. Activity C starts at the end of A and B, so the arrows representing A and B have to be bent in some way such as that tried in Fig. 2.5(b), to bring their ends back together; C is now drawn on from this new junction, and all is well—so far. But when we come to insert the arrow for activity D, we hit a problem: D must fit on the end of B, but this now also makes it look as though D can only start after *both* A and B finish, which is not the intention at all. This difficulty will always arise where the network branches, and the branches split, and then partially re-join across the division between the main branches.

(a)

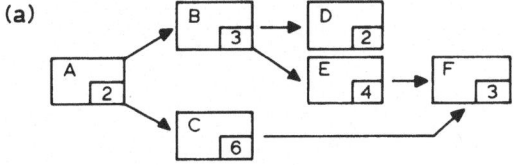

(b)

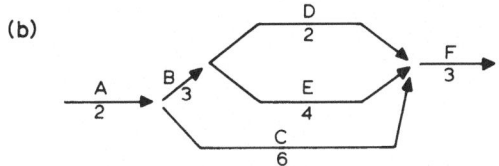

(c)

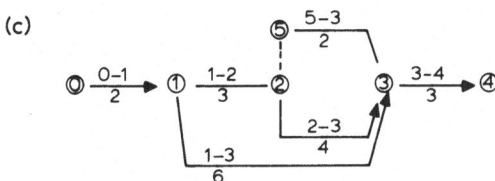

Fig. 2.4. Activity-on-arrow notation. (a) Network from Fig. 2.2(b) in activity-on-node format; (b) same network, redrawn in activity-on-arrow format; (c) alternative method of naming activities.

The way round the difficulty is to insert a dummy activity, which takes no time and uses no resources, between the end of B and the beginning of C, as shown in Fig. 2.5(c): now, clearly, starting C depends on the completion of both A and B, but starting D depends only on the completion of B, and the succeeding dummy.

Unfortunately, it is not easy—even for people familiar with this method of presentation—to get the positioning of dummies right all the time. Worse than this, this method makes it much more difficult for non-specialists to read or draw up networks. During the data collection phase (discussed in the next chapter), drawing a network during the discussion provides both a record of the information collected, and an aid to the analysis of the problem. If the providers of the information cannot interpret—and therefore cannot correct—the draft network, this valuable aid is lost: you can no longer prompt the discussion to move forward by

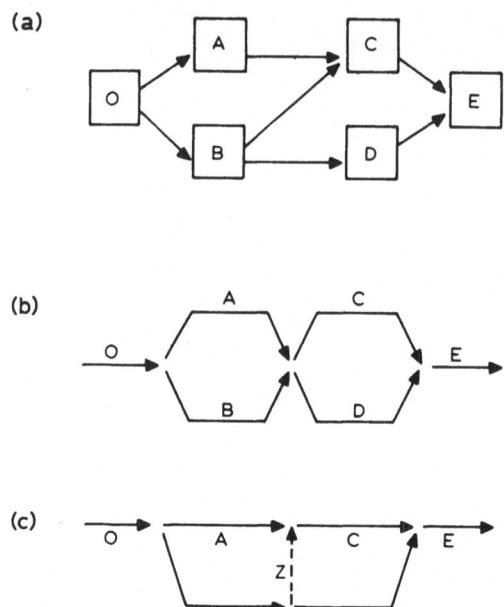

Fig. 2.5. Dummy activities in activity-on-arrow notation (durations omitted).
(a) Correct form of network in activity-on-node format; (b) incorrect conversion
to activity-on-arrow format; (c) correct conversion, using a dummy activity Z.

saying things like, 'Well, if I draw it like this, is that what you mean,
or does it omit anything important?' Experience suggests that the
introduction of dummies really does produce enough difficulty in interpretation to convert the people who provide the information from
participants in the networking exercise to mere onlookers, and this is a
significant loss.

2.10.2. Precedence Diagrams
Unfortunately, there is some ambiguity over this name: it is sometimes
treated as synonymous with activity-on-node networks, but is also used
for a method of representing the situation which occurs where successive
activities must each lag slightly behind the start of their predecessors. We
have already met with a simple example of this, embedded within our
dairy project example: laying water pipe must lag slightly behind trenching, and backfilling must lag slightly behind pipelaying. In both the

activity-on-node and activity-on-arrow formats, this situation is represented by a ladder-like portion of the network (Fig. 2.6(a)) with a rather arbitrary number of steps; the version of precedence diagramming shown in Fig. 2.6(c) is an easier way of drawing the same thing. For hand calculation, the eventual conversion of this piece of the network into a 'ladder' is unavoidable; some computer programs will accept a numerical description of a situation such as that shown in Fig. 2.6(c) directly.

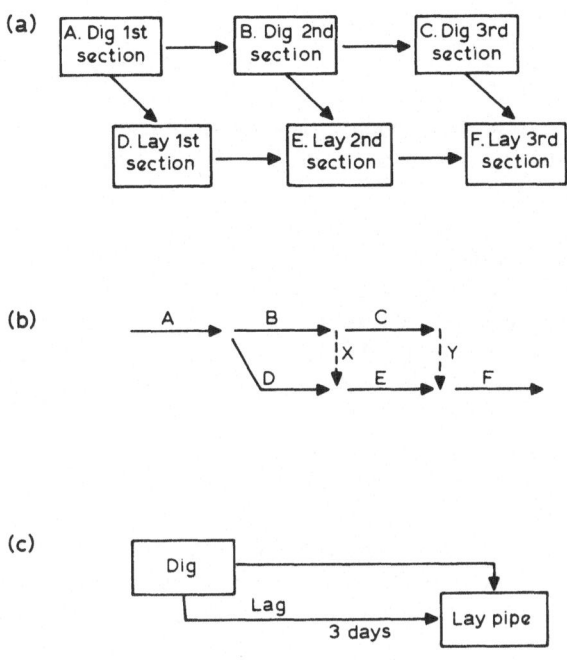

Fig. 2.6. Ladder networks: precedence diagram. (a) Activity-on-node version; (b) activity-on-arrow version, with dummies X and Y; (c) precedence diagram.

This sort of precedence diagram is also called a special precedence diagram; and the ladder format is also sometimes described as 'series and parallel', by analogy with electrical networks.

2.10.3. Event Networks
Events are things like 'importation of cows completed', 'installation of water supply started', and so on. An event networking scheme usually

represents a project by the end-points of activities; this may be compressed into a network showing only those events that conclude major groups of activities. This approach has two disadvantages; it is difficult to construct an unambiguous event network in some cases; and, much more serious, the whole approach is wrong. The idea that control is being exercised by setting up a series of target dates and concentrating only on whether these dates are met throws away far too much potentially useful warning information which is available to users of the more detailed approach advocated here. It may be a convenient summary reporting format—it certainly is no more than this.

Major events, used in this way, as progress markers, are sometimes referred to as 'milestones'—but even here there is ambiguity because the same word is also occasionally used for scheduled dates, i.e. externally imposed dates by which parts of the project must be complete (see Section 2.2.3(d)).

2.10.4. Gantt Charts

This is, in fact, simply our time-scaled bar-chart, with the logical interdependencies indicated as in Fig. 1.3. It is re-introduced here, to emphasise that the graphical method of solution is potentially a very powerful one; the problem is that it requires a reasonably skilled hand and eye throughout, and there is no way that the construction of Gantt charts can be reduced to a clerical routine, as can the arithmetic method of solution (whether it is sensible to use clerks for this processing is another matter).

Where skilled mid-level staff are available, the graphical approach is very attractive. The network can actually be maintained by the people responsible for the activities. Lowe[1] describes in detail how this has been done in practice.

2.11. NOTE ON TERMINOLOGY

In this book, CPA stands for critical path analysis (either in general, or a specific analysis of one project); CPM for critical path methods.

Other authors use CPM as either a general term (as here), as the name of a specific computer program, or to distinguish an approach to critical path methodology that is directed towards time/cost trade-offs of projects, and deliberately ignores the variable nature of activity duration.

Where this distinction is drawn, the contrasting element is usually PERT, an approach in which considerable emphasis is placed on the variable nature of activity durations, an aspect of the scheduling problem which is discussed in Chapter 6.

However, PERT is also used as a general term for critical path methods (the letters originally stood for Project Evaluation and Review Task, an assignment connected with managing the development of the Polaris missile), and has been taken over as the name of a specific computer program.

'Logic' and 'structure' are often used, interchangeably, for the structure of the network. This sort of confusion over names—seen already with 'milestone' and 'precedence'—seems to have arisen because various groups, not in full contact, developed the methodology semi-independently, at about the same time; Moder and Phillips[2] describe the development of the subject.

REFERENCES

1. Lowe, C. W. (1969). *Critical Path Analysis by Bar-Chart*, Business Books, London.
2. Moder, J. J. and Phillips, C. R. (1970). *Project Management with CPM and PERT*, Chapter 1, Van Nostrand–Reinhold, New York.

CHAPTER 3

The Network: Collecting the Data

3.1. INTRODUCTION

In the previous chapter, we looked in some detail at the method of solving the basic problem of project control: how to set targets for progress that are reasonable, in terms of the durations of the various activities, and their inter-relationships. Then, we took the data regarding these durations and relationships as given; now, we must look carefully at how such data are obtained, since the quality of the results will depend very largely on how well this is done. In the course of this, we will discover that many of the items of input—e.g. project duration—require considerable care in their definition.

To try to make the discussion both more concrete, and more widely applicable, we will define the context as that of a large agricultural project, being constructed for a government agency, which will be referred to as the client. This hypothetical agency has appointed a project management consultant, who is providing the project manager (PM). On his staff is the individual who is responsible for the CPA of the project, and whom we will refer to as the analyst. There will also be the agencies responsible for the conduct of the various parts of the implementation, such as contractors, or even sections of the client's own organisation; these will be referred to as 'sections', and their respective heads, as 'section managers'. Beneath these, there will be persons responsible for groups of operations within sections—e.g. within the direct works group in our example, there will be a foreman responsible for the fencing work. These we will refer to as technical managers.

Obviously, there are all sorts of variations on this structure: the PM may be a member of the client's staff; or several of the functions may be fused in one person, e.g. the project manager may also be the analyst.

Indeed, on a small project, he may also be all the heads of sections, and all the technical managers as well; and section head and technical manager will very often be the same man. Nevertheless, all the functions are there, and will all have to be dealt with. It will probably have struck the reader already that the workability of any project control system depends very much on the formal structure of the relationships between all these parties. Without the requirement to provide feedback to (or to respond to advice from) the PM being built into the formal relationships between the sections and the client, the project manager will be left in the position of having neither information on progress, nor control over events. How this situation is prevented from arising is described in Chapter 7.

It is strongly recommended that the collection of the data follows the scheme:

(a) an initial session with heads of sections, at which the analyst outlines what he is doing, and why, and gets a general picture of the project;

(b) filling in the detail at a series of meetings with technical managers and heads of sections where appropriate—the specialist sessions;

(c) a confirmatory meeting with heads of sections, to review the network.

This system ensures that the people whose cooperation is being sought know why they are being asked to devote time to the exercise, and provides a general framework of knowledge which the analyst builds on and finally checks.

3.2. THE INITIAL SESSION

The initial session should be held with the project manager, and the heads of all the various sections under his control; e.g. on an irrigation project, these might be the contractor's senior technical man, the land reclamation specialist and the agricultural manager. It is quite important to call all these into a single meeting, not so much to save the analyst from having to repeat the same presentation, but to establish a style of openness of communication which is essential to obtaining information about a system—the project—most of whose component parts affect each other at some stage.

There are two schools of thought regarding the presence of the client's staff at this first meeting: in many cases, where the project manager is actually a member of the client's staff, there is no choice. In others—as in the case of our hypothetical project—where the PM is on the staff of a consultant firm retained by the client, this choice does exist: although, theoretically, the client is represented by his agent (the PM) the greater informality of this arrangement and the fact that comments can be made off-the-record can be a distinct asset to the analyst. This is especially true where parts of the project have been contracted out to agencies who, in the words of one analyst 'got the job and then figured out how to do it':[1] in the more formal circumstances of a meeting with the client, such a contractor will provide no information which might be used against him, and might choose to give no information whatsoever.

This initial session could well be organised along the following lines.

3.2.1. The Analyst's Introduction

Here, perhaps, the title of this chapter is not entirely appropriate: the first object should be, not 'collecting data', but disseminating information.

Very often, when a CP analyst enters a project, it is as a result of a relatively high level decision: since he will necessarily be taking an interest in all aspects of the project, the decision that he should be involved is, almost inevitably, taken at a level senior to that of the various heads of sections. This can cause problems: the analyst may be regarded either as the client's spy on the long-suffering section heads (especially by contractors), or he may be the focus of resentment if he appears to be interfering in their various areas of responsibility.

There are two ways in which the CP man can be introduced into a project. The first of these is by far the best: this is where either the project manager, or a member of his staff, is the analyst, and has a recognised part in the project structure from the outset. Conversely, the other is by far the worst: this is where the analyst is brought in at a late stage of a disorganised or over-running project, to attempt to salvage it. In this case, the fact that many irretrievable errors in organising the project have already been made is not the worst problem; nor is the fact that the situation will continue to deteriorate during the period in which project staff and the analyst are getting to grips with it. His real problem is that almost everyone on the project will regard him as being there primarily to assign blame, and will adjust the information they feed him accordingly.

Whichever of these cases applies, the analyst has first of all to make the following things clear:

(a) he is not exercising a technical function: that is, he will be taking the advice of the section heads and technical managers on how things are done (although he may need to question the consistency of their statements);

(b) he will not be in any way cutting across the established lines of responsibility, but feeding advice into the existing command structure;

(c) he is not there to apportion blame for delays and disruption, but to help keep the project on schedule (although he should be honest and admit that CPA will inevitably result in greater accountability in this respect than most other project management systems);

(d) what sort of information he requires, and why: this should consist of a very brief (e.g. five minutes) outline of the CPM approach.

Not only should the analyst make these things clear, he should act on them: the worst and commonest failing is in relation to item (b); that is, in failing to act through the existing command structure. This is a great temptation to an outside 'monitoring' consultant appointed by, say, an aid agency.

The final part of the introduction will almost always consist of some sort of reply to the objections that are usually put forward to attempts to organise a project in this way: these are usually presented in the form of a claim that in this particular situation (country, or type of project, or whatever else), the reliability of estimates of durations of activities is particularly low. We will see why this objection is not valid in Chapter 8—basically, it is because CPM provides the basis for a good system of running adjustments to counteract the effect of unforeseen delays.

3.2.2. Getting the Picture

The analyst's next job, after making his introduction, is to gather the following items of information, in order to establish the overall pattern of the project.

(a) What are the main disciplines involved? This will usually be obvious from the job designations of the participants at the meeting, but will include things like: land clearance; irrigation works; land reclamation; building; agriculture. At this stage, the analyst should ensure that the formal structure of the project is clear in his mind. This is particularly necessary for relationships with contractors and sub-contractors, which will become important if the question of reorganising working methods to meet deadlines comes up (see Section 5.3).

(b) What are the major activities needed to complete the project? We will be returning later to the problems of defining activities (see Section 3.3), but at this stage we can say that something represents a major activity if it is the responsibility of one of the section heads, or one of his subordinate technical managers, and is relatively independent of other parts of the project. 'Relatively independent' means that, although it will have to follow or precede at least one other major activity, these will link with it either at its beginning, or at its end, but not onto some point inside it. An example will make this clear: bush clearing on a large project will not usually be one major activity, even though it may be the sole responsibility of a specialist unit, because other sections' activities may link into the middle of it, e.g. road making or primary rough ploughing could start on one block while clearing proceeds on another. However, bush clearing *on one block* could well be defined as a single major activity: it has to be completed before anything else (ploughing, fencing, putting in water supplies, etc.) can proceed on that block; it has to follow other jobs, e.g. setting out, and getting the clearing machinery onto the site; and it is completely independent of some other jobs, e.g. building houses elsewhere on the scheme. This sort of information is best collected at a joint meeting of the section heads, because it is much more likely that the points at which the major specialities interact with each other will be spotted. By 'interact' we mean link up, in such a way that some of the work of one section cannot proceed until some part of the work of another is complete. The reasons why this is interesting should be obvious from the preceding chapter: each interaction will represent a node in the network.

At this stage, it will normally be fairly clear what order the jobs can be done in, and an attempt should be made to get the participants to assist the analyst to finalise this information in the form of a network, of the type described in Section 2.2. This sort of visual recording of the information, is, in practice, a very good method of stimulating thought about the relationships of project components among people whose normal responsibility is for the technical quality of a single component.

(c) How long will the major activities take? Unless the section head has substantial experience of similar projects in a comparable area, this sort of off-the-cuff reply may not be very reliable. However, it will still be useful in revealing whether or not there is a conflict, within a section, on what are reasonable rates of work.

(d) Who is responsible for each activity, in the sense of having day-to-day control of the work, i.e. who is the technical manager, as we have defined it? It is essential that the analyst eventually establishes some sort of common agreement between section heads and technical managers on work rates, especially if there is any conflict with the rough estimate referred to in (c). If this is not done, the end point of any query on delays during implementation will often be the remark 'well, if you'd asked me, I'd have told you that wasn't possible...'

3.2.3. Format of the Initial Session

A good format for the meeting is that the analyst should chair it. He should work through the items in Section 3.2.1 and 3.2.2; then sort the major activities into a rough order of execution; then, taking them in turn, ask each of the section heads for the points at which his discipline interacts with the one currently under consideration. Physical maps and plans are obviously very useful at this stage: wherever there is a physical intersection, there is the possibility of an interaction. This stage should end with the production of a sketch for the skeleton of the network. On large and complex projects, it may be preferable to go into the questions of 'How long does it take?' and 'Who is responsible on a day-to-day basis?' later with the individual section heads, to avoid the meeting dragging on and some participants losing interest.

3.3. SPECIALIST SESSIONS

These are normally individual meetings with the technical specialists responsible for the major activities, as identified at the previous stage. These meetings have the following functions: dividing the major activities into activities; determining the relationships between them, and finding out how long they are expected to take, and what resources they will require.

3.3.1. The Activities

Definition
At this point, it is necessary to define the word 'activity' more carefully than we have done so far. Up to now, the word has been used as though it was roughly the same as 'job', 'undertaking', or similar vague words. In CPA it has a very tightly defined meaning, in terms of the following

properties:

(a) *Homogeneity*: First of all, an activity must be a physically homogeneous process in the sense that you can realistically ask, how far is this physically complete? You can ask this of bush clearing, when the answer will be something like, 'twenty per cent of the area is done', or excavating a drain of uniform cross-section, when the reply might be 'half the length is done'. It's not sensible to ask this sort of question of, say, installation of a milk processing plant, because, except in financial terms, there is no way of adding together site preparation, building the shell, providing services, installing all the various items of equipment and so on. (If you still feel that percentage financial completion is a useful figure, re-read Section 1.3!) You will now appreciate that our example is unrealistic in this respect. It was made so, to reduce the bulk of the calculations, which depends largely on the number of activities in the network. (However, for a given number of activities, the number of convergences and divergences also affects the amount of calculation.)

The reason for this part of the definition is that, at a later stage, it will be necessary to obtain and use realistic measures of the rate of physical progress, in controlling the project (see Chapter 8).

The question of scale comes in here: there will always be some sort of limitation on the time and other resources available to the analyst, and he cannot handle too many activities. Where there are large numbers of identical units—e.g. grain storage silos—the activity might be defined as 'building 36 silos', and this is homogeneous in the sense defined provided that each one is always completed as a unit, by the same method. If this is so, you can ask, 'How much physical progress has been made?' and the reply, 'Half of the silos have been completed' is clear and unambiguous. If the work is being done by an experienced contractor, his estimate for the time required for each unit could be used; if not, a CP analysis of the building of one silo could be made, and the result taken as the unit time.

In the special case of administrative activities, the proportion completed can really only take one of two values: 0% and 100%, i.e. a project is either approved or it is not.

(b) *Absence of interaction*: The activity must only link with other activities at its beginning and end. For example, 'building 36 silos' cannot be treated as a single activity if two of the silos are required to test-run an associated processing plant, in such a way that the activity 'test-run' links in part way through 'build silos'. This part of the definition is there because each such linkage represents a node in the network; it is possible that, for example, the test-run and the construction of the first two silos could be on the critical path, while the construction of the remaining silos was not. It would be impossible to handle the calculations unless the completion of the first two units was separated from the remainder.

(c) *Independence of sub-activities*: The activity must only be capable of being done in one way (apart from trivial variations in the order in which sub-activities are done). In the case of the silos, for example, if it is possible to work on one or more silos simultaneously, then building the group cannot normally be treated as a single activity. This is because the time to be assigned to the job will depend on how many units are being worked on at once, and this is a question that has to be determined at the resource scheduling stage. If, however, there is only one specialist crew and they can only do one silo at a time, then building the group becomes one activity.

(d) *Presence of defined end-points*: Activities must be things with clearly defined end-points. If this condition is not included, bits of work simply get left out. For example, the boundaries of activities in a sequence such as the following must be carefully defined:

Major activity	Activities
Get funds released:	PM prepares schedule of expenditure and sends to client department.
	Client department scrutinises, and prepares draft administrative approval for expenditure.
	Finance Department compares draft with budget and gives final sanction to Treasury to release funds.

A sensible definition is that each activity ends when the papers have been typed, signed, and actually delivered to the office responsible for the next activity. If this precaution is not taken, it is quite easy for the analyst to come up with a total time for the sequence which excludes typing, despatch and postal times, and in some organisations this fraction can equal or exceed the time taken by the actual work.

If all the parts of the definition are observed, it may seem, at first, to produce an excessive number of activities to handle. However, this multiplicity has the advantage of helping ensure that nothing is left out, and focusing the technical manager's attention on all the steps that have to be completed. This is important in getting realistic replies to the last part of the problem—the timings. It also has the advantage that it becomes possible to estimate the extent of progress on major activities like 'get funds released' from the number of the component activities that have been completed.

3.3.2. Relationships between Activities

This is the second thing to be determined at the specialist sessions. As we have seen, CPM works with three forms of relationship between any pair of activities: they can be completely independent; one may necessarily follow another; or one may necessarily precede another. These are all the possibilities; out of these the more complex relationships described in Section 2.2 can be built.

Generally, it is not too difficult to get information about the interrelationships of activities, provided that the analyst ensures that the technical manager realises that the analyst doesn't want to know what order things are usually done in, or what order it is currently proposed to do them in. What he needs to know is what restrictions physical and logical necessity place on the order that they must be done in.

3.3.3. Timings

The third question which the specialist sessions should provide an answer to is, 'How long will each activity take?' This is a much thornier problem, at least in the context of projects in developing countries, with inexperienced staff, and often with inexperienced contractors.

It is a continual surprise, particularly in the case of contractors, that physical and financial calculations can be done in considerable detail, and with great thought, while times and work rates are almost totally ignored. Frequently this results in enormous delays relative to the contract period, and often it invalidates those same calculations, which depend on the time the contractor expects to be working for.

Whenever possible, the analyst should try to start with asking the work rate for particular activities, rather than for completion time. He should probe along one or other of the following lines, to attempt to check the validity of the estimates:

(a) previous experience—how long did it take on a similar job on another project, with a similar workforce?
(b) job analysis—is it possible to break the job down into components for which there is some basis for comparison? For example, this agency may never have built a lined watercourse before, but they

know how long it takes for excavation, simple formwork, and laying concrete slabs.

Sometimes, neither of these is possible, and the figure is little more than guesswork; where this happens, the forecasts of completion dates produced by the management information system (see Chapter 8) will tend to fluctuate wildly in the first few recording sessions, and then stabilise as experience accumulates on the job. The worst situation is where there is a major activity, late in the project, whose duration is very uncertain; in this case, there really is no answer, and even in technologically developed cultures, this situation usually results in both cost and time estimates being exceeded. This is no reason to fail to manage those parts of the project that are manageable.

It is important that the technical manager understands that by 'How long does it take?' the analyst wants to know how long this activity is likely to take, in the actual circumstances of the project, *not* how long it 'should' take. Often, things can be done a lot faster than they are, but it is the typical time that is wanted. For example, if senior staff appointments have to be scrutinised by some form of public services commission, and this body takes three months to do what could be done in three days, three months is the period the analyst must use. The reason for this is that whether the delay is due to pressure of other work in the commission, or mere inertia, these won't change merely because the project has come into existence. Only if the project manager has the control required to ensure that things are done at greater speed should you use the time in which things 'should' be done.

The analyst must also remember that the work rates, particularly for physical (as distinct from administrative) activities always depend on the method which it is proposed to use. However, provided that activities are defined as suggested above, much ambiguity is avoided: the contention that 'Well, it depends how many men you put on it' disappears if the consideration that it can only be done in one way is enforced. That way is then recorded—see below.

One point is usually very difficult to put over to technical managers: while some of them are optimistic, and some are pessimistic, a few believe they are clever enough to beat the system by specifying unreasonably long times for the activities under their charge, thereby taking the pressure off themselves. This is a false hope, because there is a good chance that they will only focus attention on themselves by making their activities appear to lie on the critical path.

3.3.4. Format of Meetings

These will normally be just an informal two-man discussion. However, it is worth recording the information formally, on a pro forma such as that shown in Fig. 3.1. The meaning of the first three column headings will be obvious from what has already been said. The one labelled 'method' is provided to enable the analyst to enter a brief comment on the technology to be used; for example, in the case of contour terraces for soil erosion control, whether they are to be hand-excavated, cut with a grader, or dug out by the blade of a bulldozer. This column is intended to tie together the two halves of the form—the second half consists of a list of the main resources required for the activity and its use will become clear when we have looked at resource scheduling. Although it might look tidier to complete this second half of the form at the meeting at which the network is established, in practice this is nearly impossible: especially if

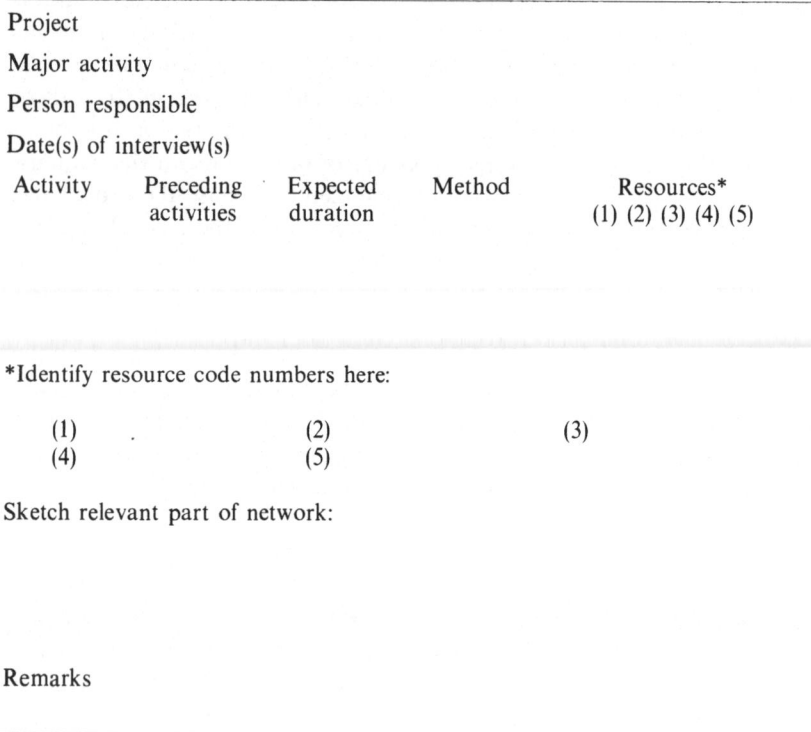

Project

Major activity

Person responsible

Date(s) of interview(s)

Activity	Preceding activities	Expected duration	Method	Resources* (1) (2) (3) (4) (5)

*Identify resource code numbers here:

(1)   .   (2)   (3)
(4)   (5)

Sketch relevant part of network:

Remarks

Fig. 3.1.   Network data recording form.

Table 3.1. Check List of Activities

(A) If the activity currently under consideration involves a physical item of work, are any of the following activities needed:
   1. Getting any sanctions required (e.g. import permits, permits to install fuel tanks, customs clearances).
   2. Provision of funds.
   3. Tendering/purchasing/hiring.
   4. Transporting to site.
   5. Provision of access (e.g. roads, delaying other activities that could make access difficult).
   6. Getting equipment (e.g. fencing tools, bush clearing machinery).
   7. Providing personnel required to do installation/work.
   8. Providing temporary/permanent storage/housing for the item.
   9. Site preparation/setting out/clearing/foundations or other fixings.
   10. Provision of services: water, power, fuel, spares, repairs and maintenance, fertilisers, seeds.
   11. Provision of operating staff: managers, clerical staff, farm workers.
   12. Testing/commissioning/clearing up.

(B) If the activity involves personnel, are any of the following activities needed:
   1. Getting necessary sanctions for establishing the post(s), for engaging personnel, bringing in expatriate personnel, transfer of local staff.
   2. Provision of funds.
   3. Advertisement, selection and other recruitment procedures.
   4. Provision of housing and transport.
   5. Training.

(C) If the activity is an administrative one (e.g. issuing a permit) are any of the following activities needed:
   1. Preparation of application.
   2. Endorsement of application (e.g. import permits may need a certificate that the item is unobtainable locally; applications for confirmation of senior appointments may need to be endorsed by a senior official).
   3. Issuing formal notices (e.g. applications for planning permission).
   4. Arranging committee meetings.

---

the technical manager is unfamiliar with project planning techniques— which is almost always the case—the analyst will have enough problems in getting the network information sorted out to exhaust both himself and his colleague. The space at the foot of the form is provided for a sketch of the relevant part of the network; the value of getting the technical manager to participate in drawing this cannot be over-emphasised. Needless to say, it is wise to do this is pencil!

It is very useful if the analyst can apply a check list to each activity, to ensure that all the associated activites are included. Such a check-list rapidly becomes part of the analyst's mental equipment, so that there is no question of having to run through a written list at each activity. However, at some stage, the list does have to be formalised, and the one shown in Table 3.1 is recommended. The use of the check list will often generate queries relating to other technical managers' areas of responsibility, and these should be recorded systematically in the 'remarks' panel of the form.

Often, such queries will refer to administrative activities—issuing permits, releasing funds, sanctioning the engagement of staff, and so on—and it is important that the analyst understands the administrative system he is working with. Few things are more frustrating on a project than to discover, too late, that a major delay has occurred, because an essential sanction was not obtained; one of the few things which *are* worse, is to discover that such a delay has resulted from an attempt to get such a sanction by the wrong method, or from the wrong body. Real life project networks, except for very minor projects, typically do contain a substantial number of administrative activities, particularly in more bureaucratic cultures.

3.4. THE NEXT STEP

In practice, the less project staff know of project management techniques, the more repeat visits to the technical managers will be required to resolve queries. The only consolation the writer can offer is that this is fairly normal!

At the end of the specialist sessions, the analyst should have all the information required to draw the network, and to get on with the calculations described in the last chapter. That phase of the work concludes with a provisional presentation of the results of the network analysis (Section 2.9); the next step is to discover whether or not the first schedule of starting dates requires further adjustment to take account of resource constraints, and that is the subject we must look at next.

REFERENCE

1. Aarkays Associates (1980). *Guidelines on Installing PERT CPM Systems*, Aarkays Associates, Karachi.

CHAPTER 4

Resource Scheduling

4.1. DEFINITION OF SCHEDULING

It might seem that the schedule of starting and finishing dates for all the project activities produced at the end of the initial network analysis (Section 2.6) provides the schedule according to which the project should be run. This is far from true: no account was taken, during the forward and backward pass calculations, of the possibility that activities might compete for scarce resources. For example, the activities of the direct labour force, as originally scheduled (Table 2.3) require road making, fencing, and the installation of part of the water supply to go on simultaneously. Now, these activities will certainly all require some of the same resources, such as site transport or unskilled labour. It may be that there are not enough of these resources available to enable all these activities to go on simultaneously. If this is the case, some adjustments will have to be made to the provisional timetable.

The process of making these adjustments is called resource scheduling, or more briefly, scheduling. Its end product is a schedule of starting and finishing dates which is feasible, both in terms of the logical structure of the project—what has to be done, and in what order—and of the limitations on available resources. The original timetable (Table 2.3) is only certain to be feasible if there are no such limitations. Obviously, in practice, this will be a rare event.

4.2. DETERMINATION OF RESOURCE REQUIREMENTS OF ACTIVITIES

4.2.1. Identifying the Relevant Resources

Usually, it is said that durations of activities should be estimated without regard to the resources available. This is true in as far as different activities are concerned: the technical managers should not be attempting to adjust the durations of the various activities because they may compete for resources. For example, if both road making and soil conservation works require surveyors for setting out, the technical manager should not be stretching his estimate of the time required to allow for the fact that these activities may compete for surveyors: until the network and CPA is complete, there is no guarantee that the two activities would be in progress at the same time, and the adjusted estimates would therefore be unduly pessimistic.

However, it is obvious that in the case of any one activity, the duration depends very much on the method and resources used—the rate of progress of say, bush clearing, will depend very much on the type of equipment used and this must be noted. Provision was made for this in the pro forma recommended in Chapter 3 (Fig. 3.1).

Also, the resources available can affect the shape of the network: suppose, for example, that 2000 metres of drain are to be cut, and that it is operationally convenient to do all the setting out at once. If only one machine (e.g. a dragline) is available for the work, the relevant piece of the network will look like Fig. 4.1(a): a straight sequence of setting out, followed by excavation. If, however, two machines are available, the network will be modified as in Fig. 4.1(b), with a branched sequence, in which excavation of the first and second portions of the drain both start simultaneously, at the end of setting out.

Before we can look in any detail at the estimation of resource requirements, we have to decide which resources are relevant. At first sight, it might appear that the number of possible categories is enormous—all the grades of manpower, and all the different items of machinery and equipment. However this huge and daunting list, far too long to be easily processed in the way to be described below, can be drastically pruned by taking account of the following considerations. We have seen that the reason for doing this part of the exercise at all is that different activities compete for the same resources. Therefore, only those resources that are likely to be competed for need be considered. Prominent among these will usually be road transport, site transport, unskilled labour, skilled labour needed for a variety of jobs (such as masons and carpenters), and equipment usable for a variety of work (such as bulldozers). All such resources need to be considered during scheduling. However, some resources are physically incapable of being used for more than one activity. A canal lining machine is a good example: its work rate determines the duration of its activity, and that is

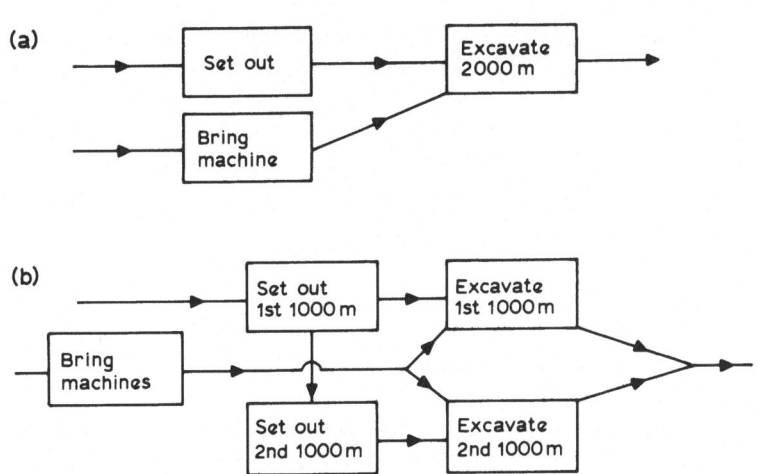

Fig. 4.1. Example of the effect on the network of different availabilities of resources. (a) One machine available; (b) two machines available.

that. Similarly, some resources are limited by administrative decision to only one type of work: for example a specialist mechanised unit for soil conservation work may be provided, with the proviso that its equipment cannot be borrowed for other types of work. All such resources can safely be ignored during scheduling of other activities.

The major categories of resources which will be involved in the scheduling exercise are those that are involved in a wide range of activities, and the chief of these are:

(a) Funds: especially where a project runs over a number of accounting periods (usually financial years), the proportion of the funds that will be available in any one period may have been determined without much reference to the phasing of physical activity. For example, the determining factors may have been the availability of an aid agency's funds, or the government's budgetary limitations. The effects of this will have to be evaluated during scheduling. Of course, scheduling should be capable of producing a very strong case for adjustments to the financial allocations for the various periods of the project's life.

(b) Professional administrative time: while the most senior men will usually not be engaged in routine tasks, it is entirely possible that shortages of middle management staff will mean that only one man is available to undertake key tasks, such as interviewing prospective senior staff or evaluating tenders. Alternatively, only one man may have the seniority necessary to liaise effectively with other agencies; often, it will require intervention by a very senior man to expedite things like an order for equipment which has to be processed by a different government agency, for example.

(c) Skilled staff: e.g. surveyors, for setting out soil conservation works, roads, etc.; masons, for building watercourse linings; carpenters, and so on.

(d) General labour.

(e) General purpose equipment: road transport; or site transport (e.g. tractors and trailers); concrete mixing plant, if concrete is required for several different activities.

(f) Non-physical resources: these comprise items which are really channels, rather than things; e.g. there may only be a certain quota of cement, steel or fertiliser allowed into a country in a particular year, and the project may only get a fixed share of that amount.

With these considerations in view, it may be possible to cut the number of resources to be considered drastically. Exactly which resources should be included can best be discovered by discussions at section head level.

4.2.2. Collecting the Data
This will require a repeat visit to the technical manager to complete the right-hand section of the pro forma shown in Fig. 3.1. The only point requiring particular mention is that the analyst needs to ensure that the duration previously estimated is consistent with the amounts of resources now claimed to be necessary; this is where the 'method' column on the pro forma comes in: it acts as a reminder to both people involved in the discussion of what was the original basis of the estimate of duration.

4.3. DETERMINING THE AVAILABILITY
OF RESOURCES

Determining some of the limits may be relatively straightforward: in the case of government projects, funds, professional and skilled personnel and most physical items will have been budgeted for in great detail and the problem may be that the limitations are far too clearly defined. They may be determined so rigidly that it is very difficult to make essential adjustments.

The biggest problem is usually general labour: where this is being drawn from the locality, even the total number who will come forward is probably unknown, and their tendency to leave the site at harvest times, or festivals, or work short hours during customary fasting periods will be likewise an unknown quantity. This can make a policy of importing the labour force from an adjacent urban area (or even another country) attractive: such labour has at least no local commitments of its own, and is a known quantity. Often, however, this is not possible because of the high cost of importing labour and providing accommodation, or because of social or political problems with strangers in the area. Also, there may be a need to get local participation in the project. Where these considerations make it necessary to use labour from the project area, and where there is any doubt about the adequacy of the labour force, it is worth taking some form of simple census:

— How many adult workers live within reasonable travelling distance?
— How many of these have permanent jobs?
— How fully occupied are they with work at harvest times?

This obviously requires some effort—but, on a big project, the quantity (and quality) of labour can be a serious problem, and deserves careful evaluation.

A point which is often overlooked is that the constraint, i.e. the amount of a particular resource which is available, may not be constant over the period of the project. The chief examples of this are general labour, and funds. In the former case, there may be festivals which will, in effect, reduce the amount of labour available. In any case it is unwise to plan the labour use of a project in such a way that the number of workers engaged for the duration of the project shoots up at the beginning, stays high, and then declines abruptly at the end. Such a pattern makes no allowance for the initial difficulties of recruitment and getting men accustomed to the work, and ignores the fact that the labour force, knowing the end of the job is in sight, may begin to drift off to other work before the project is completed. (This, obviously, is only relevant where paid labouring is an established occupation.) The case of funds is slightly more complicated; the best way to handle the cash flow of a project is as a cumulative sum. The requirement of activities therefore becomes a series of upward steps, each representing either a lump sum of capital expenditure, or regular payment of recurrent costs, such as salaries, office rents, and so on. The constraint—usually, the

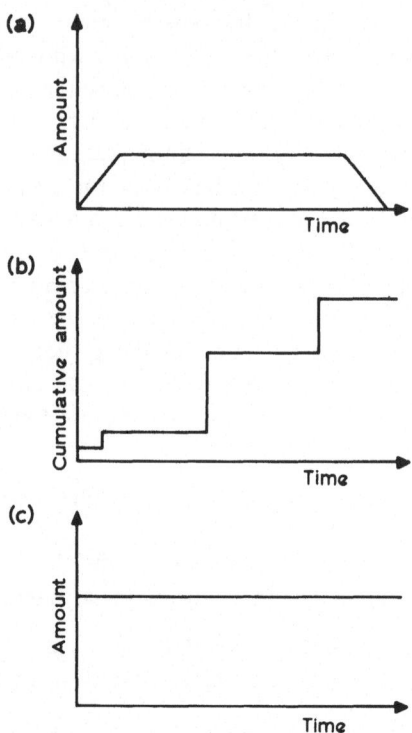

Fig. 4.2. Some patterns of resource constraints: (a) ideal pattern for labour usage; (b) typical pattern for funds; (c) fixed constraint.

amount set aside in the budget—when expressed in cumulative terms, will have the same general shape. Most constraints including permanent staff, will, however, take the form of a constant limit throughout the project; the three patterns of constraint are shown graphically in Fig. 4.2.

4.4. ASSESSING THE RESOURCE REQUIREMENTS OF THE INITIAL SCHEDULE

The initial schedule is the set of early starts and finishes, which results from the initial network analysis described in Chapter 2, and summarised in Table 2.3. Early starts (and finishes) are used because any delay in beginning an activity after its earliest possible start means that some of the safety margin of time (in the form of float) for that activity and its successors has already been consumed, with no gain.

The easiest way to handle these results is to convert them back to bar-chart form, by exactly the same method that was used for the watercourse improvement project in Chapter 1. This has been done in Fig. 4.3 for our sample dairy project. The only differences are: the bars have been made deeper, so that the resource requirements of each activity can be written on them; the activities have been blocked into segments, i.e. the continuous sequences of activities lying between convergences and divergences in the network. The reason for this difference is that it is usually a good idea to move all the activities in a segment together, when trying to even out total resource requirements. Because there is no free float within a segment, if one of the earlier activities is moved, the later ones must also be shifted; also, there is a lot to be said during the actual work for keeping up the momentum on a particular technical group of activities, if only to ensure that the work force is not puzzled by irregular stops and starts, which tend to reduce their confidence in the management (which is often already low). It is not, however, essential to move segments as a unit, and if unavoidable, even activities may be split; many computer programs for resource scheduling do just that (see Section 4.8).

In our example, we are assuming that the other resources which the preliminary enquiries identified as being likely to limit the rate of progress are: the project manager's time, as spent on specific tasks, such as tender evaluation and negotiation; general semi-skilled labour within the direct works units, for tasks involved in roadmaking, fencing, laying on water supplies; and tractor/trailer units for site transport and in the case of tractors, powering a rear-mounted digger. The requirements have been written on the bars of the chart; an attempt has been made to vary them in a realistic way (e.g. in the case of the livestock tenders, it has been assumed that the manager is an agriculturist, and time has been allowed for him to travel to inspect and select the animals). The chart also shows the free float for each segment, and the total float. By holding a straight-edge vertically on the chart and sliding it across, a week at a time, the chart can be viewed as a series of columns, each containing the requirements for each resource for the relevant week. These are totalled, and the sums transferred to the top section of Fig. 4.4 (considered later). (It is easier to ensure that the changes in resource requirements between activities in the same segment are not missed if alternate activities are coloured or shaded in some way.)

It is immediately obvious that the initial timetable is not going to work, because it requires more of the resources than are available, at least for certain peak periods; however, the amounts of float on some of

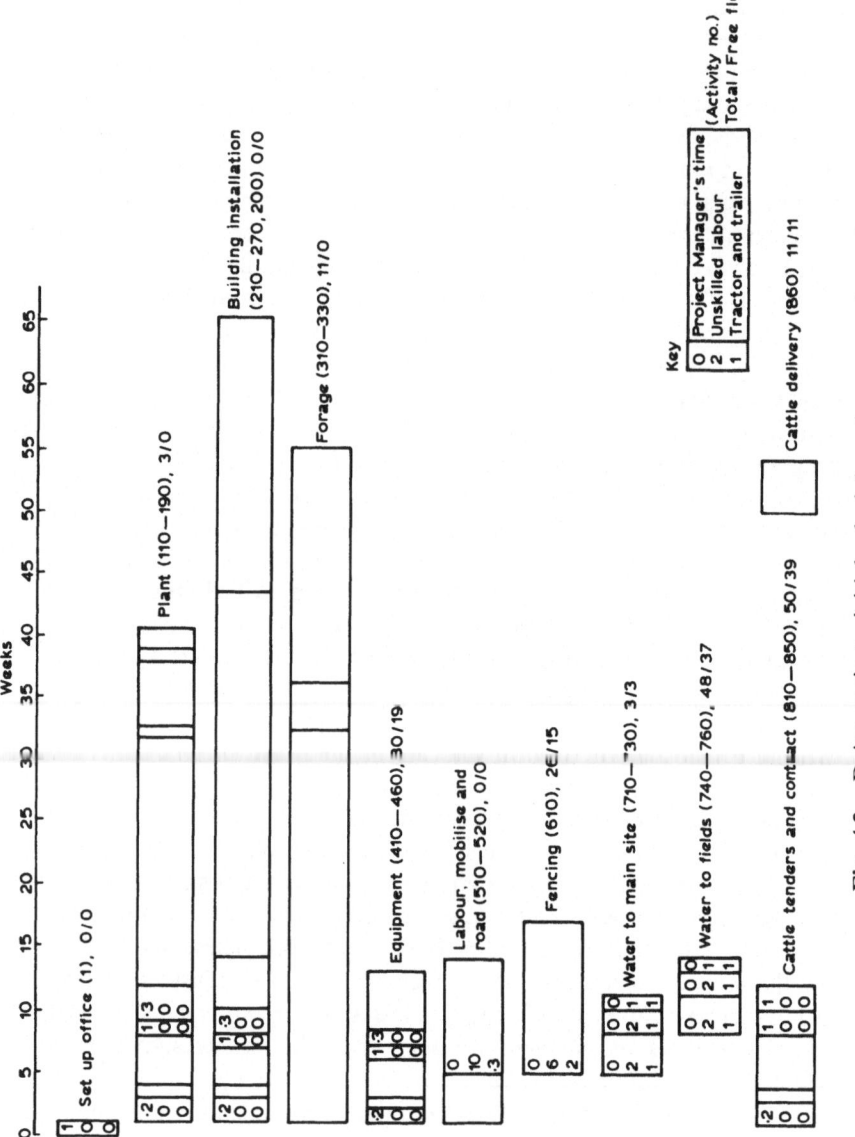

Fig. 4.3. Dairy project—initial schedule and resource demands.

the resource-using activities give plenty of scope for adjustment. Before seeing how the adjustment is made, it is worth looking at two aspects of the chart. The first is that, even with this relatively small increase in complexity, compared with the first example (Fig. 1.3), the chart has already become a little difficult to read. For example, there should be an arrow linking the eighth bar into the middle of the third bar, representing the necessity for a water supply to be available at the main site (activity 730) before building (270) can begin. For anyone not familiar with the network, it would not be easy to see which (if any) other activities this arrow would connect with, and it would tend to give the impression that there are extra sub-divisions within the bars it cuts through. This all emphasises a point made earlier: complex networks are not easy to handle graphically, simply because of the possibility of visual confusion. (They also lack all the internal cross-checks described earlier.)

The other aspect of the chart worth mentioning is that the relationship between total and free float shows up clearly: on the ninth bar (the segment containing the activities concerned with providing a water supply to the fields), each activity has a total float of 48 days. However, the free float is only 37 days. This is shown as the distance between the end of the segment and a vertical line (not marked) representing the latest finishing of the group of segments that have to be completed before the cattle can be brought in. If the start of the first activity in the field water supply segment is delayed by 48 days, that vertical line will have to be pushed to the right by 11 days; it will carry with it the lower-most bar, the segment representing delivery of cattle. The 11 days is the interfering float on the ninth bar/segment; and all the difference between the ninth segment's total float and its free float does turn up, downstream in the network, as free float, as you should have expected. The important point is, if it is necessary to move a segment by more than its free float, the timing of other segments is also affected: failure to allow for this will produce a schedule that will not work, because it will attempt to complete activities in the wrong order. But a segment can be moved by an amount less than, or equal to, its free float, with absolute confidence that this cannot cause problems elsewhere.

4.5. RESOURCE LEVELLING

This means attempting to level out excessive peaks in the graph of demand for the resources to bring them within the constraints, with the minimum effect (preferably none) on project completion time.

4.5.1. Choosing the Order in which Resources are to be Levelled

Normally, the best method for making this choice is simply on the basis of cost, that is, unit cost of the resource. There are two reasons for this. First, as the demand for successive resources are levelled, jobs requiring them tend to be pushed further and further apart, and further and further downstream in the project; the resources which are levelled last are therefore rather more likely to be left idle for short spells, so that the cheaper the unused resource is, the less this idle time costs. Secondly, as the resource levelling process continues, and float is used up, the project management may eventually be faced with the choice between increased project duration, and purchasing more of certain resources; if the resources are ranked for processing in order of decreasing cost, the resource which has to be purchased will be one of the cheaper ones. Some resources may not have a price, merely a fixed amount that is available; these have to be given high priority. (An economist, on the basis that the cost of an item is really what you have to forego to get it, would say this is quite consistent, because you would have to forego an infinite amount of other things to get something which is not available at all.)

4.5.2. Choosing the Order in which Activities are to be Levelled

It should be made clear at this point that the system we are looking at concentrates on one resource at a time; when adjustments have been made to level the demand on it, these are not usually disturbed without the most pressing reasons. If this restriction is not imposed it is quite easy to build new peaks of demand for the first levelled resource, in adjusting the demand for later ones. What we will be doing is shuffling the activities about within the available float, to level out the peaks of the resource load. Clearly, on occasion, there will be a choice of activities that can be shuffled, and some method of establishing priority is required. It is suggested that the activities competing for the resource currently under consideration at any one date (on the chart) are ranked according to the following list. The highest ranked activities get the resource; when there is none left, the lower ranked activities are delayed. The order of priority is:

(1) *Lowest total float*: this means that critical activities will always be moved last, and not at all if this is avoidable. In the case of non-critical activities, giving priority for allocation of resources to those with least float is justified by the fact that float represents a margin of safety to cover delays in work on the activity and errors arising from genuine uncertainty over how long the activity will

actually take: giving priority for resource allocation to activities with lowest total float means moving activities with greater total float first, thus helping to keep these safety margins evenly distributed over project activities, as far as possible. Activities which rank equally on total float can be separated on the basis of the second criterion, which follows.

(2) *Greatest work content for the resource*: those activities which make a smaller demand on the resource should be the ones that are moved, and fitted into periods when odd spare quantities of the resource are available. Activities which rank equally on both float and work content can be separated by the third criterion, which follows.

(3) *Late finish date*: this is reputed to work well in practice, on the basis that activities with a late finish date, which falls in the latter part of the project period, tend to suffer more from the effects of accumulated delays and disasters. Therefore, it is alleged, delaying them in the schedule will fit better with what actually will happen. This is little more than a tie-breaker, and if any activities still rank equally after all these criteria have been applied, then an arbitrary tie-breaker must be applied: the serial numbers of the activities are as good as anything, but toss a dice if you are a gambling man.

The whole process of ranking resources for levelling, and levelling unacceptable peaks of demand by moving activities with lowest priority for the resource, is a method of getting a reasonably good fit between available resources and the demands of competing activities, without unreasonable increases in project completion time. It is not, in any sense, an optimisation process: it will not produce an answer to the problem: 'What is the absolute best possible schedule?'; such methods are available,[1] but they require more calculation than the one set out here, and suffer from other disadvantages. For now, it is sufficient to say that the method described will always give a reasonably good result, and will always be practicable, requiring no more than graph paper and pencil: methods aimed at overall optimisation of the schedule usually require a computer, and many projects are implemented in circumstances where the idea of computerisation is totally unrealistic.

4.5.3. Levelling Resources on the Sample Project
We will assume that the process of discussion with section heads produced the conclusion that the only relevant resources were, in order of decreasing priority, the project manager's time, and general labour

and tractor/trailer units within the general works unit (the tractors are used for site transport of men and materials and for powering diggers for trenching for the water supply works). The requirements for these resources were extracted from the form (Fig. 3.1), and written on the bar-chart, Fig. 4.3. Looking at the summary in the top section of Fig. 4.4 (which has been reduced in size to fit the page), we see that the amount of the various resources available will be exceeded—if we attempt to implement the initial timetable—at the following times:

(1) during weeks 8–12, the demand on the project manager's time for specific, demanding tasks is too high.
(2) during weeks 6–14, the demand for general labour in the works unit is too high.
(3) also during weeks 6–14, more tractor/trailer units are required than are available.

This illustrates the point that increasing the number of resources con-sidered during the scheduling process will not necessarily cause a pro-portionate increase in the amount of work involved, since, in some cases, levelling the demand for one will automatically/help to level the demand for another.

Before we start to move the activities about, it is important to adopt a standard method to avoid certain problems. First, it is essential, as we have seen, to ensure that if an activity is moved by more than its free float the effect on other activities is properly allowed for; and, secondly, the effect of any move on all the resources must be carefully recorded, so that, at each move, the true current pattern of resource demand is being used. A layout for the calculations, which does meet this requirement, is shown in Table 4.1.

The first move shown in Table 4.1 is only partly to do with resource levelling: two of our segments are subject to the limitations that they should not end more than a certain length of time before another starts: delivery of cattle should coincide as closely as possible with plant start-up, and quotations should not be obtained more than 12 weeks before delivery starts (this being the maximum period the suppliers will fix the price for). Both these requirements can be more than met by delaying the relevant segments, as shown in the first panel of Table 4.1; this also helps to even out the demand on the project manager's time. The panel shows the calculations of the new resource demand profile which results from the move. There is still an overload on the PM, however, so, how do we remove it?

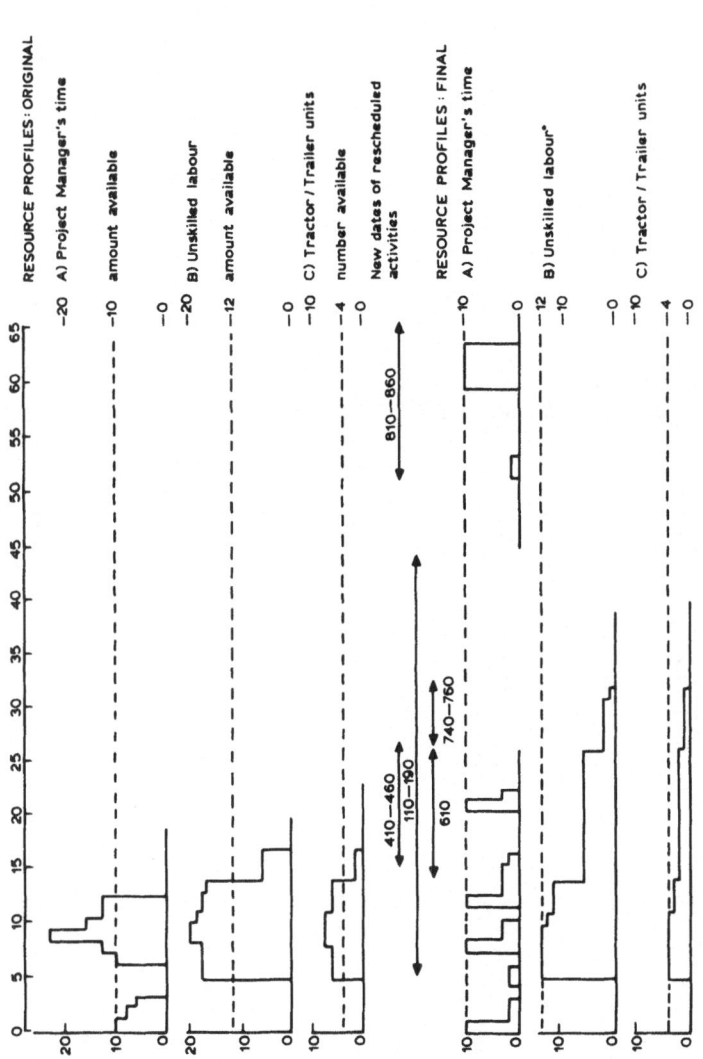

Fig. 4.4. Resource scheduling for dairy project.

Table 4.1. Resource Scheduling Calculations for Dairy Project

Step number	Activities moved	This move levels demand for	Type of resource	Resource demand (weeks)																
				1	2	3	4	5	6	7	8	9	10	11	12	13	14	15	16	17
	(Original state)		P	1	0·8	0·6				1	1·3	2·3	1·6	1·3	1·3					
			U						18	18	18	20	20	19	18	18	17	6	6	6
			T						6	6	6	7	7	7	6	6	6	2	2	2
1	810–860 (cattle)	P	P'		(0·2)	(0·2)				(1)		(1)	(1)	(1)	(1)					
			P	1	0·6	0·4				1	1·3	1·3	0·6	0·3	0·3					
			U						18	18	18	20	20	19	18	18	17	6	6	6
			T						6	6	6	7	7	7	6	6	6	2	2	2
2	410–460 (equipment)	P	P'	(0·2)						(1)	(0·3)								X	X
			P	1	0·4	0·4				1	1	1·3	0·6	0·3	0·3					
			U						18	18	18	20	20	19	18	18	17	6	6	6
			T						6	6	6	7	7	7	6	6	6	2	2	2
3	110–190 (plant)	P	P'	(0·2)	(0·2)			0·2	X	(1)		(1)	(0·3)	(0·3)	1		0·3	0·3	X	X
			P	1	0·2	0·2		0·2	0·2		1	1·3	0·3	0·3	1	0·3	0·3	0·3		
			U						18	18	18	20	20	19	18	18	17	6	6	6
			T						6	6	6	7	7	7	6	6	6	2	2	2
4	610 (fencing)	U and T	U'					X	X	X	X	X	X	X	1	X	X	X	X	X
			T'															(2)	(2)	(2)
			P	1	0·2	0·2		0·2	(6)	(6)	(6)	(6)	(6)	(6)	(6)	(6)	(6)	(6)	(6)	(6)
			U						(2)	(2)	(2)	(2)	(2)	(2)	(2)	(2)	(2)	(2)	(2)	(2)
			T					X	2	2	2	2	2	2	2	2	2	2	X	X
5	740–760 (water to fields)		P	1	0·2	0·2		0·2	0·2		1	0·3	0·3	0·3	0·3	0·3	0·3	0·3	0·3	0·3
			U						12	12	12	14	14	13	12	11	10	6	6	6
			T						4	4	4	5	5	5	4	4	3	2	2	2
			U'							(2)	(2)	(2)	(2)	(2)	(2)	(2)	(1)			
			T'							(1)	(1)	(1)	(1)	(1)	(1)	(1)	(1)			

Step number	Activities moved	This more levels demand for	Type of resource	18	19	20	21	22	23	24	25	26	27	28	29	30	31	32	33	34
	(Original state)		P																	
			U																	
			T																	
1	810–860 (cattle)	P	P'																	
2	410–460 (equipment)	P	P'	X	X	X	1	0·3	X	X	X	X	X							
3	110–190 (plant)	P	P'	X	X	X	X	X	X	X	X	X	X	X	X	X	X	X	X	X
			P				1	0·3												
			U																	
			T																	
4	610 (fencing)	U and T	U'	6	6	6	6	6	6	6	6	6								
			T'	2	2	2	2	2	2	2	2	2								
				X	X	X	X	X	X	X	X	X								
			P	6	6	6	1	0·3	6	6	6	6								
			U	6	6	6	6	6	6	6	6	6								
			T	2	2	2	2	2	2	2	2	2								
5	740–760 (water to fields)		U'										2	2	2	2	2	1		
			T'										1	1	1	1	1	1		
													X	X	X	X	X	X		
			P	6	6	6	1	0·3	6	6	6	6								
			U	6	6	6	6	6	6	6	6	6								
			T	2	2	2	2	2	2	2	2	2								

Table 4.1—*Contd.*

Step number	Activities moved	This more levels demand for	Type of resource	35	36	37	38	39	40	41	42	43	44	45	46	47	48	49	50	51
	(Original state)		P U T																	
1	810–860 (cattle)	P	P'																	
			P U T																	
2	410–460 (equipment)	P	P'																	
			P U T																	
3	110–190 (plant)	P	P' {	X	X	X	X	X	X	X	X	X	X							
			P U T																	
4	610 (fencing)	U and T	U' {																	
			T' {																	
			P U T																	
5	740–760 (water to fields)		U' T'																	
			P U T																	

Step number	Activities moved	This move levels demand for	Type of resource	Resource demand (weeks)																
				52	53	54	55	56	57	58	59	60	61	62	63	64	65	66	67	68
	(Original state)		P																	
			U																	
			T																	
1	810–860 (cattle)	P	P	0·2	0·2															
			U	X	X	X	X	X	X	X										
			T	0·2	0·2															
			P'								1	1	1	1						
			U'			X	X	X	X	X	X	X	X	X	X	X	X	X		
			T'								1	1	1	1						
2	410–460 (equipment)	P	P'	0·2	0·2															
			P								1	1	1	1						
			U																	
			T																	
3	110–190 (plant)	P	P'	0·2	0·2															
			P								1	1	1	1						
			U																	
			T																	
4	610 (fencing)	U and T	U'	0·2	0·2															
			T'	0·2	0·2															
			P								1	1	1	1						
			U																	
			T																	
5	740–760 (water to fields)		P	0·2	0·2															
			U																	
			T																	
			U'								1	1	1	1	1					
			T'								1	1	1	1	1					

P = Project managers time; U = unskilled labour; T = tractor and trailer units—each at end of preceding move.
P′, U′, T′ are changes resulting from current move (parentheses indicate deduction).   X = new dates of activities.

The overload now occurs in weeks 8 to 9 only, and three segments use PM's time during this period. In order of decreasing total float, they are the segments relating to locally purchased equipment, plant procurement, and tender processing for the building (which is critical). The total float criterion has ranked the activities without any ties and there is no need to refer to the work content and latest finish criteria. Obviously we must move the locally purchased equipment segment first, but a preliminary check shows that this alone will not completely solve the problem, and plant procurement will also have to be delayed; as the latter can only be delayed by three weeks, it makes sense to delay the other segment by enough to move its peak demand for PM's time clear of the existing peak *and* the dates at which plant procurement will draw on PM's time after it is moved. A little forethought along these lines will save a few steps (although plodding mechanically through the method will also get you to the same point eventually). So, the new position for the segment is recorded (the Xs); the resource requirements are added in at the new dates, and deducted at the old dates and the adjusted resource profile is carried forward to the next panel.

Had the move exceeded the free float on the segment, all the moves of the relevant downstream segments would have been recorded in the panel along with their consequences for resource requirements. Had it been necessary to move a critical activity, there would be no difference in the method—but project duration would have been lengthened, of course.

The successive panels of Fig. 4.4 show the final moves required to level demand on the PM's time, and the two moves that level demand on general labour and the tractor/trailer units, which introduce no new principles; the lower-most panel shows that the schedule is now within the bounds set. The new positions of the segments that have been moved can be picked up from the table, and a new bar-chart and resource profile drawn, for presentation to the project management and client.

4.6. CONSEQUENCES OF RESOURCE LEVELLING

It is important to notice that one of the consequences of making the adjustments is that some float is eliminated. In projects where the limiting resources are widely-used items (e.g. unskilled labour, transport), for which different activities are competing throughout most of the project, there can be an almost complete elimination of float. In our example, float has been eliminated in the cattle and plant segments. This means that there is much less safety margin available within the project as a whole.

If float is completely eliminated from a segment, that segment will now be on the critical path—either as a parallel critical segment, or possibly replacing the original critical path entirely.

4.7. PRESENTATION OF RESULTS

At this stage, the work should be taken back to the heads of sections and presented in this form:

(i) the bar-chart after scheduling: the salient features are the total duration and the total demand for each major resource together with its time distribution;

(ii) the bar-chart before scheduling: total duration, which, by a difference from (i), shows how resource constraints have affected duration;

(iii) the network: shows critical path, with brief summary explanation of why the critical path is the length that it is.

Possibly, at this stage, all is well: the completion time is acceptable, and the resources are within limits. If this is not the case, this stage has to be handled with care: the worst thing that can happen is that the analyst—especially if he is junior to the PM (and especially if the PM is a contractor's employee and a contract has already been signed)—is howled down because the conclusions he presents are embarrassing. His best defence—and it needs presenting tactfully—is that it is the section heads' own data that was used, and the discussion should be diverted as quickly as possible to constructive ways of solving the problem. These are the subjects of the next chapter. Obviously, they will have to involve the provision of extra resources, and this means that the client will have to be approached, for a preliminary opinion on the extent to which he is prepared to relax the original schedule, and for guidelines on the additional resources—particularly funds—which he would be prepared to make available for given decreases in project duration.

4.8. ALTERNATIVE RESOURCE SCHEDULING METHOD

An alternative method of resource scheduling works on a day-at-a-time (or week-at-a-time) basis; this method is used by some computer programs. It can produce a very fragmented schedule, as there is no guarantee that, say, the second day of an activity will immediately follow the first. Would-be users of computer programs for CPA should be aware of the existence of this method and its drawbacks.

In this method each time unit—day, week, or whatever—is taken in turn; in any one unit, the first resource is checked, to see whether the demand for it is excessive; if it is, then the resource is allocated to activities, in decreasing order of priority; those activities that get resources are scheduled to be carried out on that day; those that do not, are postponed. The process then goes on to the second, third, etc., resource. The rules used for ranking resources and assigning priority to activities are similar to those described above.

4.9. MULTI-PROJECT SCHEDULING

At first sight, it might appear that the problem facing a major agency, with a number of relatively independent projects to control, is little different from that facing the manager of a project with numerous distinct contracts on it: in principle the whole work load should be capable of being treated as one network, with relatively few events— usually called interfaces—at which the sub-networks interact with each other. For example, among the projects in the pipeline of an agricultural department, might be an irrigation project and an extension service improvement project; the former might include efforts to improve water-use management on individual farms, using personnel from the latter to teach farmers. The end of mobilisation of the extension personnel and the beginning of providing the advice under the irrigation project would be an interface between two sub-networks which otherwise could be processed independently, with an early start for the advice being passed up from the extension sub-network, and a late finish for the mobilisation being passed down from the irrigation sub-network.

The snag is that often, major projects are at very different stages: one or other project may be at such an early stage that the calculations cannot be completed, and a guess is inserted. For example, if the irrigation project was in a very early state of planning, a guess would have to be made on the late finish date for the mobilisation. This guess then tends to become an imposed date binding the irrigation project (as yet unplanned), with all the problems we have already seen with imposed dates, not least, that it may impose an impossible condition on the project when it is finally elaborated. Worse still, interfaces with other projects may look both ways, i.e. a project might require both an early start date for one activity and a late finish for another, from the same unfinished project, and more than two projects could be involved. Resource competition can present similar problems; it would be fine if projects were planned and implemented in order of priority, so that the least urgent or rewarding ones suffered from resource shortages or imposed dates resulting from earlier and therefore higher priority projects. However, in practice, the order in which projects can be planned and started can be strongly influenced by relatively trivial considerations, such as the time taken to agree terms with an aid agency, or the time taken to make the administrative changes needed to implement the project. Obviously, the date on which someone conceived the basic idea, and the relative political weights of various departments promoting projects will also have an effect.

Therefore, the multi-project scheduling problem is difficult to solve in theory; however, it is also complicated in practice by two other conditions. First, the networks will be large, and may therefore require computerisation; and, secondly, departments have neither the direct, closely focused motivation of a project manager required to implement a CPM-based management system, nor, indeed, any means of acquiring resources for it.

As a result, for the foreseeable future, it is not likely that multi-project scheduling will be an important tool in the administration of agencies such as government departments. Inter-project dependencies for events will continue to be handled by scheduling and crashing procedures to bring dates of interface events into line (in preference to using imposed dates). Inter-project competition for resources will be handled by scheduling and crashing later projects within the constraints imposed by the total pool of the relevant resources, and the drain on this of earlier projects—probably with a little bit of ad hoc re-distribution, necessitating some adjustment of the earlier projects.

REFERENCE

1. Moder, J. J. and Phillips, C. R. (1970). *Project Management with CPM and PERT*, Chapter 8, Van Nostrand–Reinhold, New York.

CHAPTER 5

Review of Working Methods

5.1. ALLOCATION OF EXTRA RESOURCES

At the end of the scheduling process, the time required to complete the project may still be unacceptably long. This means that some activities may have to be shortened; obviously, from what has been said so far, it is the critical activities that will have to be shortened and, normally, this will involve the use of additional resources.

Unless the need for this has been detected at the planning stage, it may be enormously difficult to obtain the required resources: if the CPA has been postponed until well into the implementation phase, a completion date and a set of budgets and sanctions for the supply of resources which are not consistent with each other will have been built into the project approval document. If the project's sponsoring agency is a responsive, project-oriented organisation, this may not be too serious; if it is a rigid bureaucracy, with a folklore of demotions and dismissals for infringements of the rules, getting agreement to the extra expenditure may itself take up substantial amounts of time.

Regardless of when the exercise of compressing the schedule is carried out, the technique will be the same. It is often referred to as crashing, from the analogy of a railway train being compressed against a set of buffers: the wagons correspond to the activities; the free space between their buffers corresponds to the floats; the buffer springs to those parts of the activities' durations that can be shortened by using extra resources; and the locomotive corresponds to the driving need to shorten the schedule. This analogy is developed in the next section.

5.1.1. The Mechanical Analogy
Imagine a network like that shown in Fig. 5.1, consisting of five activities.

82

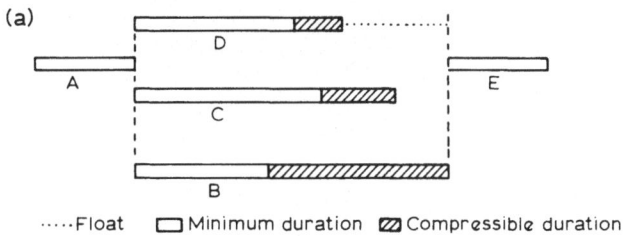

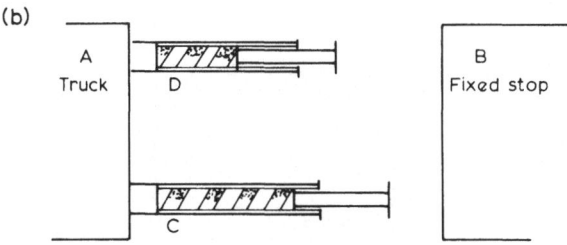

Fig. 5.1. Crashing—initial state. (a) Bar-chart of real situation; (b) mechanical analogy.

The first (A) and last (E) are start up and finish jobs (such as setting up a project office, and handing over the project to the authority which will run it); the critical path runs through them, and through activity B. If it is not possible to shorten A or E, but B can be shortened, the other segments of the network behave like the buffers on a railway wagon, approaching a fixed stop at the end of the line.

First of all, as more resources are allocated to the critical activity B, which corresponds to the locomotive pushing the truck, the free space between the trucks is used up. This free space corresponds to free float on activities C and D: reducing it costs nothing more, beyond the expenditure on shortening B.

Eventually, however, the situation shown in Fig. 5.2 is reached: one or other of the truck's buffers (we obviously have fairly badly built rolling stock, with asymmetric buffers) has contacted the fixed stop, and extra energy must be applied to bring the truck closer to the stop. This extra energy is needed to compress the spring in the buffer, in the mechanical model of the network; it corresponds to extra expenditure needed to shorten activity C, on top of that already being spent on shortening B.

(a)

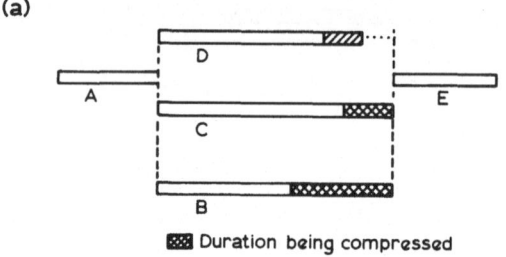

■ Duration being compressed

(b)

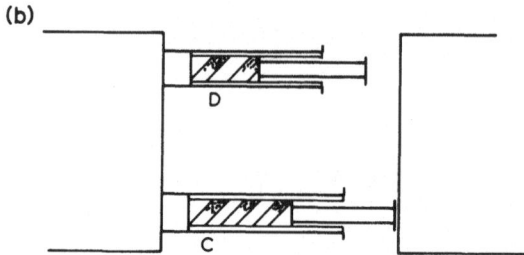

Fig. 5.2. Crashing—intermediate state. (a) Bar-chart of real situation; (b) mechanical analogy.

Further reduction in the length of B—corresponding to more pushing by the locomotive—will at some point cause the spring of the buffer D to be compressed, i.e. a further increase in expenditure will be needed to speed up activity D.

Finally, the point depicted in Fig. 5.3 is reached: one of the buffers is now firmly against its stop, and the truck has crashed solidly against the fixed stop. No amount of pushing by the locomotive will bring the truck closer to the stop; no amount of extra expenditure on B will shorten project duration. Figure 5.4 shows how project duration and expenditure on extra resources are related.

This, then, is the reason behind the name of the process—crashing. It is perhaps one with unfortunate associations, which all too often follow in practice! The mechanical analogy does, however, illustrate most of the major problems associated with attempts to shorten project duration by purchasing extra resources for critical activities:

    (a) each activity will have a 'normal' length—as originally used in the network—and some shorter length which can be reached by the

(a)

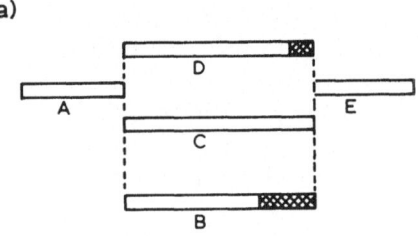

(b)

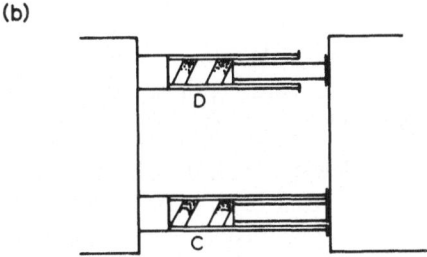

Fig. 5.3. Crashing—final position. (a) Bar-chart of real situation; (b) mechanical analogy.

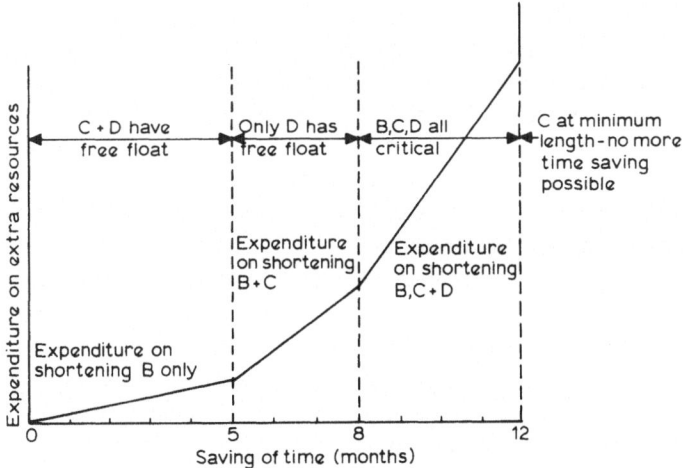

Fig. 5.4. Effect of crashing on expenditure.

expenditure of effort, i.e. the provision of additional resources; but it will always have an irreducible minimum length, below which providing extra resources has no effect;

(b) as the critical activity under consideration is shortened, other activities that converge on the same point (event) will become critical; further shortening will require expenditure on these activities, too;

(c) it will only be possible to shorten these activities by a certain amount, and, once their minimum duration has been reached, extra resources devoted to the original critical activity will have no effect;

(d) there may be a sequence of critical activities, with different costs per unit of time saved—this would correspond to a series of trucks, linked into a train, with springs of different stiffness, in the mechanical analogy.

However, there are a few crucial features of networks which the simple mechanical analogy doesn't have. First, there may not be a steady gain of time as more resources are applied; there may be only the choice of two completion dates, at different prices. Secondly, there is the effect of altering start and finish dates on all the resources included in the scheduling: shortening some activities and starting others earlier may create fresh peaks of resource requirements (see Fig. 5.5), thus undoing the good work accomplished during resource scheduling. Therefore, the cost of shortening an activity must include an element to cover buying more of any resource that would become limiting. This will mean that the cost of shortening an activity may vary—usually it will increase, the greater the shortening. Alternatively, the existence of the resource constraint will decide the minimum duration to which the activity can be shortened. (The whole process may require a couple of attempts, since often the analyst will not know what resource peaks will occur until he has tried to shorten the duration.)

Next, we have to consider the selection of activities to be crashed.

5.1.2. Priority of Activities for Crashing

The obvious criterion is cost: we would like to achieve any specified saving at the cheapest total cost. If, initially, we suppose it is desired to save eight weeks on the duration of our example dairy project, the first step would be to accumulate a table of costs per unit of time saved; hypothetical results of doing this are shown in Table 5.1. These assume

(a)

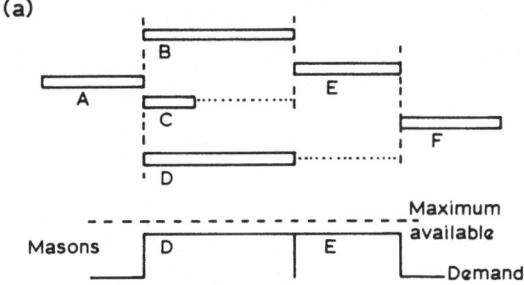

(b)

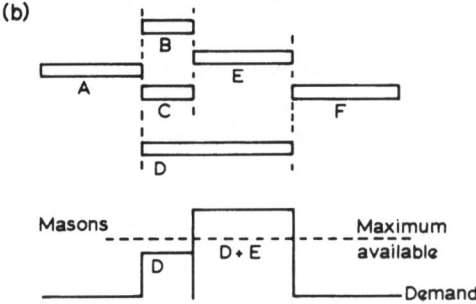

Fig. 5.5. Effect of crashing on resource demand. (a) Uncrashed—activities D and E both use masons. (b) After crashing activity B.

that the plant contractors are only willing to quote for two rates and dates, but that building can be shortened by a varying amount, depending on the amount of overtime worked.

Table 5.1. Cost of Crashing Various Individual Activities for the Dairy Project

Activity	Cost	Time saved	Cost per week saved	Remarks
Installation	Rs 40 000	2 weeks	Rs 20 000	All or nothing
Manufacture	Rs 500 000	4 weeks	Rs 125 000	All or nothing
Building	Rs 15 000	per week up to 6 weeks	Rs 15 000	

This table includes cost/time saving data for activities that are not critical, but which may become so as the durations of the activities on the current critical path are shortened.

Of the critical activities, the cheapest, in terms of cost per week saved, is building. Recalling our mechanical analogy, we have to look at the other activities that converge on the same point (corresponding to checking the amounts of free space or of compressible spring on all the buffers on the same truck). The only one is plant manufacture and delivery: if we accelerate building, project duration will not be shortened, because the next part of the critical path cannot be started until plant manufacture and delivery is complete. We also know we cannot simply slide back the start and finish dates of the plant-related activities, because they are pinned at their present dates by resource constraints (see Table 4.1): effectively, these constraints have made this segment critical.

Therefore, building and manufacturing can only be accelerated in tandem, which means there is a joint cost of Rs 125000 + Rs 15000 = Rs 140000 per week, and a joint limit of four weeks, set by the maximum time saving available on plant manufacture.

The only remaining activity on which cost can be saved is installation of the plant; the cattle-procurement segment of the network converges onto the same end-point, but this can be slid back, because we know from the final resource profile that this will not create problems (see Fig. 4.4 and Table 4.1).

We can now re-write Table 5.1 with all the entries completely independent of each other and of any cost and time interactions with other parts of the network. This is shown in Table 5.2 below.

Table 5.2. The Costs of Crashing Activities in the Dairy Project

Activity	Cost	Time saved	Cost per week saved	Remarks
1. Building + plant manufacture	Rs 560 000	4 weeks	Rs 140 000	All or nothing
2. Installation	Rs 40 000	2 weeks	Rs 20 000	All or nothing

In trying to reach our target saving, we will use these up in the obvious order, to get a total time saving of six weeks, at a cost of Rs 600 000. However, this is only half the problem: it is obviously necessary to set the costs of expediting the project against the benefits of doing so: essentially we are looking for a least-cost solution to the problem. Before going on to look at that, in the next section, a word of warning is necessary: there will always be a reason why the original working method (and the work rate that went with it) were selected. Even if the reason was only inertia, it implies that the working method selected was among the most familiar of the methods available. An attempt to substitute a novel, unfamiliar method—especially if it involves a major leap in the technical skills required—may produce a saving on paper which is never going to be reflected in actual progress. Proposals involving this sort of technical leap, as a means of expediting activities, should be scrutinised very carefully before they are included in the schedule.

5.1.3. The Least-Cost Solution

We talked, at the beginning of this chapter, about a schedule being 'unacceptably long'; it is now necessary to clarify this a little, because there is almost always some form of trade-off between duration and cost. There are three ways it can be unacceptably long, all of which boil down to the question of the cost of delay:

— there may be the danger that, if completion is delayed beyond a certain point, the work done will be seriously damaged (e.g. by seasonal weather changes in the case of major erosion control works);
— there may be cash penalties imposed from outside, if the project overruns (this would be the case where the project consists of a contractor's obligations);
— it may simply be cheaper to do it more quickly.

In the first and second cases, there is a dramatic cost increase at some fixed point in time; in the latter, a more steady increase. Take the case of the project shown in Fig. 5.4, and assume that the relevant costs are:

— overhead costs of the project administration, an equal amount in each month;
— a rapidly growing loss if a saving of at least six months is not made, this saving being required to complete part of the job before the onset of a monsoon season, and the cost representing potential deterioration.

In Fig. 5.6 these costs are superimposed on the costs of achieving the saving, and a third line shows the total cost associated with any

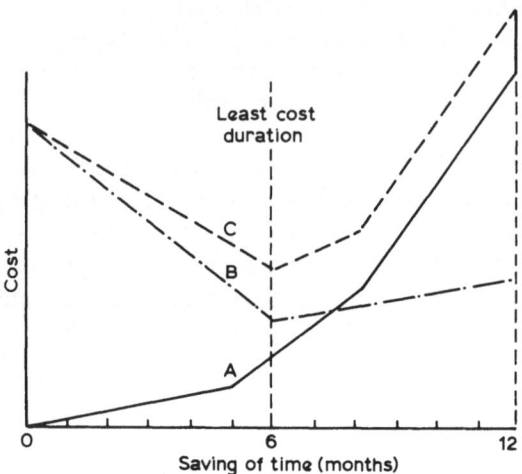

Fig. 5.6. Crashing—least-cost point. A, Cost of time saved; B, cost of delay tolerated; C, total cost.

particular level of time saving. (Remember to read the graph the right way round: project duration decreases towards the right, since it is a saving of time that is shown along the horizontal scale.)

The minimum total cost is obtained by saving six months, with greater savings than this rapidly becoming very expensive; in this case, there is a dramatic change in the cost of delay, and this is obviously the main influence in determining the position of the least-cost point. Many practical cases may not show such a sharply defined least-cost point as this: the bottom of the total cost line may be a lot flatter, and this will indicate some room for manoeuvre.

It is alleged that sophisticated contractors sometimes use this technique to decide which working practices they will adopt, and that they will include in their cost line for the cost of tolerating delay any penalties (strictly speaking, liquidated damages, see Section 7.5.1) that result from allowing the contract to overrun. This means that delays may be part of a deliberate least-cost strategy. This has implications for the design of the contractual relationships between sponsor and contractor, which is discussed in Chapter 7. In practice, on agricultural development projects, such strategies are unlikely to be met with, the real problem originating from the fact that the contractor may have bid for the job without even working out what methods he was going to use—with a corresponding vagueness on work rates and true durations. (One contractor told the author: 'I knew the job would take two years, not nine months; they knew it too—but because the Ministry said it had to be done in nine months, well, I contracted to do it in nine, and we re-negotiate the contract from time to time'!)

This least-cost procedure—starting from a reasonable schedule, and improving it—will not necessarily produce the best possible schedule, since there may be radical alterations of the basic phasing of the project that it will not explore.

5.2. ADJUSTING THE SPECIFICATIONS OF THE JOB

It may happen that either there are no activities that can be accelerated by incurring extra cost, or the cost of doing so is too great. In that case, alternative solutions will have to be sought. No solution to the problem of reducing project duration can be completely cost-free, because you really get nothing for nothing, at least in the material world. However, there are methods that may involve considerably less expenditure than crashing typically does.

One method of doing this is by adjusting the specifications of the project: for example, if building field staff housing on an extension project is a critical step, and its segment of the critical path is considerably longer than the next most nearly critical segment, the alternative of reducing the standard of housing, using local materials and labour, instead of expensive imported components, could be considered. Or, if a seed processing plant is under consideration and a technically attractive feature—say, reinforced concrete silos—are similarly making an undue contribution to the length of the CP, it may be possible to amend the design. This will always be considerably easier if the need for the change is seen at the planning stage.

Such modifications may actually result in a reduced cost. However, whether the cost is increased or decreased, once the project is under way, there will always be the same two problems. First, there may well be contractual complications: if the activity in question has been let to a contractor, he will be very wary of anything that is likely to affect his profit margins. If the total value of the job is reduced, he will feel his overheads are disproportionate to his profit; if the amount of work is increased, he will be in a position to strike a very advantageous bargain for the additional items. Secondly, it may take a considerable time to get sanction for the changes: we have already come across this problem where increases in funds are involved, and similar problems may easily occur when changes in specifications are involved. This is because financial approval documents will normally contain some indication of precisely what the money is to be spent on: if the project management tries to vary the work specification, and pay for different items, the people controlling the lowest tier of the money disbursement system will obstruct payment. This is not bloodymindedness: they have no way of discriminating unauthorised payments made in the public interest from fraudulent payments, except by reference to the work specifications, as reproduced in the financial sanctions.

5.3. CHANGES IN WORKING METHODS

If it is not possible to crash activities at an acceptable cost (or not at all), and there is no scope for changing the specifications of the work, there are other possibilities that may be considered, including the following:

(a) Some—usually fairly small—savings can be obtained by speeding up the process of contracting-out work, by either negotiating only with a selected contractor, rather than competitive tendering, or by using a leading rates contract, rather than the normal Bill of Quantities type of measurement contract (see Section 7.5.1). The latter requires considerable preparation, in estimating the quantities of all the various types of work that may be needed; in the latter, contractors are asked to submit prices for only the major items, with rates for the others being negotiated during implementation. Both these measures are really only suited to emergency situations, and may incur hidden costs, in terms of bids which have not been constrained by competition.

(b) If the trouble lies in administrative procedures, it may be possible to systematically short-cut all of these: many countries make provision for the setting up of semi-autonomous bodies, for this reason. These normally have much greater freedom to make changes in their expenditure, hire and fire staff, pay incentives for speed and quality of work, and so on. The disadvantage of this approach is that setting up such a body itself takes time, and it may not be a viable option unless the need for it is identified at a very early planning stage.

(c) Alterations in physical working practices may be considered. Most readers will have already thought of the possibility of using time and motion study. Unfortunately, this really only comes into its own on long production runs, because it is based on close study of existing, established working practices. It might be worthwhile on parts of projects where a large number of repetitive operations are involved (e.g. the making of prefabricated canal outlets). The techniques are simple in concept, but require considerable experience for successful application.

5.4. COSTING SAVINGS ON MAJOR ALTERATIONS OF SCHEDULE

In the cases looked at so far, the comparison is quite straightforward, because the time savings are fairly small. When this is not so, a complication occurs: it is likely that the net costs of the two alternatives will have a different time pattern. Typically the costs of expediting the project will occur at a very early stage, while the costs of simply allowing the delay to occur may be spread forward over several years. The extra funds tied up in the expedited project will have a cost, in terms of the other undertakings they could have financed; obviously, the simplest way of valuing their loss is by the relevant interest rate in the country.

In such a case, the best method of making the comparison is to correct the expenditure for the different dates on which it occurs, to a common base date—usually the first year of the project. (Most readers with an orthodox planning background will be familiar with the methods of doing this.) Suppose you had to compare the value of a cost of Rs 1 000 000 now, with Rs 1 000 000 to be spent in five-years time: the question you should ask yourself is, how much now is exactly worth Rs 1 000 000 in five years—to which the answer is, the sum that would grow to Rs 1 000 000 over five years, if invested. Factors by which the future sum is to be multiplied to reduce it to an equivalent value at the start of the project have been tabulated for different interest rates and time periods, and a sample of these 'discount factors' is given in Table 5.3. The process of applying them is known as discounting, and the result of the calculation, as the discounted cash flow (DCF). The appropriate rate of interest is that which the client organisation usually requires as a minimum for investment projects, since this represents the minimum value of the alternative uses of the funds.

Table 5.3. Discount Factors for Calculating the Present Value of Future (Irregular) Cash Flows at Various Interest Rates

Percentage

Year	5%	6%	7%	8%	9%	10%	11%	12%	13%	14%	15%	16%	18%	20%	25%	30%
1	0·952	0·943	0·935	0·926	0·917	0·909	0·901	0·893	0·885	0·877	0·870	0·862	0·847	0·833	0·800	0·769
2	0·907	0·890	0·873	0·857	0·842	0·826	0·812	0·797	0·783	0·769	0·756	0·743	0·718	0·694	0·640	0·592
3	0·864	0·840	0·816	0·794	0·772	0·751	0·731	0·712	0·693	0·675	0·658	0·641	0·609	0·579	0·512	0·455
4	0·823	0·792	0·763	0·735	0·708	0·683	0·659	0·636	0·613	0·592	0·572	0·552	0·516	0·482	0·410	0·350
5	0·784	0·747	0·713	0·681	0·650	0·621	0·593	0·567	0·543	0·519	0·497	0·476	0·437	0·402	0·328	0·269
6	0·746	0·705	0·666	0·630	0·596	0·564	0·535	0·507	0·480	0·456	0·432	0·410	0·370	0·335	0·262	0·207
7	0·711	0·665	0·623	0·583	0·547	0·513	0·482	0·452	0·425	0·400	0·376	0·354	0·314	0·279	0·210	0·159
8	0·677	0·627	0·582	0·540	0·502	0·467	0·434	0·404	0·376	0·351	0·327	0·305	0·266	0·233	0·168	0·123
9	0·645	0·592	0·544	0·500	0·460	0·424	0·391	0·361	0·333	0·308	0·284	0·263	0·225	0·194	0·134	0·094
10	0·614	0·558	0·508	0·463	0·422	0·386	0·352	0·322	0·295	0·270	0·247	0·227	0·191	0·162	0·107	0·073
11	0·585	0·527	0·475	0·429	0·388	0·350	0·317	0·287	0·261	0·237	0·215	0·195	0·162	0·135	0·086	0·056
12	0·557	0·497	0·444	0·397	0·356	0·319	0·286	0·257	0·231	0·208	0·187	0·168	0·137	0·112	0·069	0·043
13	0·530	0·469	0·415	0·368	0·326	0·290	0·258	0·229	0·204	0·182	0·163	0·145	0·116	0·093	0·055	0·033
14	0·505	0·442	0·388	0·340	0·299	0·263	0·232	0·205	0·181	0·160	0·141	0·125	0·098	0·078	0·044	0·025
15	0·481	0·417	0·362	0·315	0·275	0·239	0·209	0·183	0·160	0·140	0·123	0·108	0·084	0·065	0·035	0·020
20	0·377	0·321	0·258	0·215	0·178	0·149	0·124	0·104	0·087	0·073	0·061	0·051	0·037	0·026	0·012	0·005
25	0·295	0·233	0·184	0·146	0·116	0·092	0·074	0·059	0·047	0·038	0·030	0·024	0·016	0·010	0·004	0·001
30	0·231	0·174	0·131	0·099	0·075	0·057	0·044	0·033	0·026	0·020	0·015	0·012	0·007	0·004	0·001	—

Reproduced by kind permission of Professor J. Nix, Wye College, University of London.

Both streams of net costs are discounted back to the same date, when they can be directly compared. Here is an example of the method, for a hypothetical project, for which the relevant interest rate is 15%:

Year	A Project delayed			B Project expedited		
	Net cost	Discount factor	Discounted sum	Net cost	Discount factor	Discounted cost
1	500 000	0.87	435 000	2 500 000	0.87	2 175 000
2	1 000 000	0.76	760 000	600 000	0.76	456 000
3	1 000 000	0.66	660 000	—	0.66	—
4	500 000	0.57	285 000	—	0.57	—
5	500 000	0.50	250 000	—	0.50	—
Discounted net cost			2 390 000			2 631 000

Two points need to be noted: the discounting has been done on the assumption that the expenditure effectively occurs at the end of each year, so that the first year's expenditure itself has to be discounted, to calculate its value at day 1 of the project, and no interest cost figures appear, because this is taken care of by the discount factors themselves.

Now this calculation is completed, the two different streams of net costs can be fairly compared, and it will be seen that the option of allowing the delay to occur is actually cheaper, despite the fact that the total undiscounted net cost of this option is 13% higher than that of expediting the project.

5.5. CONVERSION TO CALENDAR DATES

At the end of the crashing stage (if it is needed), it is finally possible to convert the schedule into calendar dates. This is fairly straightforward, being a matter of numbering the working days consecutively (i.e. ignoring weekends and holidays) and transferring the number of days of work from the schedule.

However, a few points need care:

(a) The durations of some activities may have been given inclusive of non-working days. This is particularly likely to be true of administrative activities; if someone says 'It typically takes 10 weeks to get approval of a proposed appointment', he almost certainly means 10 calendar weeks, not 70 working days.

(b) Some activities may not stop for normal weekend rest days; anything involving expensive plant is likely to come into this category, and, of course, paying overtime for weekend working is one way of crashing an activity.

(c) The work rate on some activities may depend on the season of the year in which it falls: house building may take a lot longer in the wet season, for example. If, after the whole scheduling process has been completed, it is found that an activity of this type falls in an unfavourable season, it may be necessary to make adjustments 'downstream' from the activity affected, after increasing its duration.

CHAPTER 6

Risk and Uncertainty in Project Scheduling

6.1. INTRODUCTION

This chapter looks at three cases in which the uncertainty associated with many of the estimates which are used in CPA are of special importance:

— the uncertainty surrounding the estimates of activity durations;
— the uncertainty as to when the accumulated experience of actual activity durations begins to justify a general revision of the estimated durations, either over the whole project, or over some particular group of activities;
— the uncertainty as to the actual outcome of some activities (e.g. whether the process of establishing a post will succeed or fail).

These cases are dealt with in Sections 6.3, 6.4 and 6.5 respectively, after developing some essential statistical ideas in Section 6.2, which the reader with a good grasp of statistics may decide to skim through.

6.2. STATISTICS

The domain of statistics is quantities—such as estimates of duration and work rates—which are stochastic. 'Stochastic' implies two things about a quantity (for which we will adopt statistical terminology, and use the word variable):

(a) On the relevant scale its variation is as significant as its typical value. Scale is the crux of the matter: if an activity typically takes 20 weeks, and the likely variation is of the order of a day, the duration can safely be regarded as a fixed, constant quantity; if, however, it typically takes 10 days, but with a commonly observed range of 8 to 15 days, it may be necessary to take account of this range, i.e. to treat the duration as a stochastic variable. So far, we have assumed that activity durations are constants.

(b) The causes of the variation are many, individually small, and, collectively, beyond effective control; typical causes which would fit into this description are:

— interruptions by bad weather;
— variations in materials, in the widest sense, from building sand requiring extra sieving to variations in the size and number of large trees in bush clearing;
— variations in the amount of port congestion (affecting imported items only);
— printing delays which affect advertisements for tenders and appointments;
— length of the queue of potential appointees awaiting vetting and approval by public services commissions.

In dealing with stochastic variables, statistics uses the concept of probability, which is related to uncertainty: our uncertainty is that we do not know where in the span of possible outcomes (a range of estimated durations, for example) the actual outcome will lie; the risk we incur is that our actions may have bad consequences if the actual outcome is too far from our estimate; and probability is, in some sense, a measure of the likelihood of the various degrees of deviation of actual outcome from estimate.

Statistics mostly operates with three processes: digesting data, testing ideas about data, and predicting outcomes; and these processes operate on the main features of variables, which are their typical values and the variations about them. As a basis for the more practical applications, this section will discuss probability, the statistical processes, and the characteristic features of variables in them.

6.2.1. Probability

It is possible—easy, even, if one is so inclined—to get into philosophical difficulties over the definition of probability. For our purposes, it is sufficient to say that we expect the future to go on rather like the past, and that the probability of something happening in the future is measured by the relative frequency with which it has occurred in the past. Relative frequency could be measured as a percentage; for various reasons—chiefly mathematical convenience—it is actually measured as a proportion, so that the scale of probability runs from 0 to 1:

— something which always happens, i.e. on 100% of trials, has a probability of 1 (and there can be no greater probability than this, since nothing can occur more often than always);
— something which occurs, in the long run, in 40% of trials, has a probability of 0·4;
— something which never occurs, i.e. in 0% of trials, has a probability of 0 (and there can be no lesser—negative—probability than this, since nothing can occur less often than never).

The probability of an event can be established in three main ways:

(a) Calculation: if we want to know the probability of a fairly thrown dice showing a particular face—say, the deuce—simple arithmetic suggests it must be one-sixth (0·167), since the probability that one or other face will show is 1, and the faces have, by definition, equal chances of showing.

(b) Observation: where it is possible to observe the same piece of work (say, clearing a hectare of bush, under similar conditions) on a number of occasions, we could estimate the probability that the time taken would lie within some specified range as the fraction of the total number of observations on which the actual time taken was within that range, and use that probability in calculations about the future.

(c) Subjective estimation: this is the most difficult area, because two superficially similar ideas, very different in principle, are often confused. If a knowledgeable observer is asked to review mentally his experience of similar jobs, and estimate what proportion took, say more than 30% longer than average, the result is likely to be a useful indication of the true proportion. By an extension of this, if he is asked to visualise a number of repeats of a situation, novel but similar to others in his range of experience, and make a similar estimate, that may also be a useful estimate of the relevant probability. There may be objections to its poor accuracy, but not to the method. Although the practical usefulness of such estimates may be severely limited by their vagueness, provided that the observer is still within his sphere of competence, and has sufficient grasp of the concepts of probability to appreciate the question, then the estimates may still be usable. The problem with subjective probability occurs when an observer is being asked to quantify his state of mind, in terms of his certainty regarding something on which he has no experience; this is a very different process—but it does not affect our use of the subjective approach.

Some information on the manipulation of probabilities is necessary:

(a) The probability of two events both happening is the product of their individual probabilities: e.g. the probability of a dice showing 6 on two successive throws is

$$P_{(6)} \times P_{(6)} = \tfrac{1}{6} \times \tfrac{1}{6} = 0 \cdot 028$$

(b) The probability of either (and possibly both) of two events happening is the sum of their probabilities: e.g. the probability of a dice showing either a 5 or a 6 is

$$P_{(5)} + P_{(6)} = \tfrac{1}{6} + \tfrac{1}{6} = 0 \cdot 33$$

So far, we have talked in terms of getting the probability of an activity taking longer than a specified time. Very often we will have (or need) rather more than this: an estimate of the probabilities of the duration falling within various segments of the range of possible durations. This is called a probability distribution; an example is shown, in histogram form in Fig. 6.1(b); note that the probabilities are proportional to the areas of the bars and only incidentally to their heights. The figure also shows how the probability distribution relates to a frequency distribution (Fig. 6.1(a)).

Probability distributions—like probabilities of the individual happenings that make them up—can be generated by calculation, observation or subjective estimation. Often, in fact, attempting to get a subjective estimate of the probability distribution of an activity duration will clarify the respondent's ideas, since he will have to eliminate any inconsistencies in his ideas regarding the relative probabilities of different parts of the range.

The general shape of the probability distribution shown is significant: variables subject to stochastic variations arising from many individually small influences tend to have this bell-shaped probability distribution, regardless of the shapes of the probability distributions of the individual disturbing influences themselves. As more and more influences affect the criterion, as more repeats of the sampling process are made, and as the width of the bars of the histogram are reduced, the graph tends towards

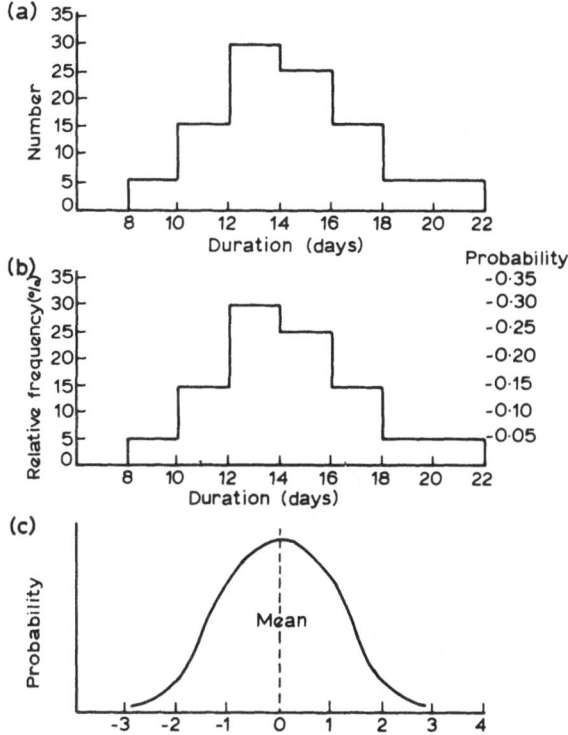

Fig. 6.1. Probability distribution and normal curve. (a) Frequency distribution of 100 durations—areas of bars proportional to frequency. (b) Probability distribution—areas of bars proportional to relative frequency (%) ≈ probability. (c) Normal curve—horizontal scale in units of SD—scale shifted, so mean = 0.

the smooth, bell-shaped curve known as the 'normal curve' (Fig. 6.1(c)); it is possible to demonstrate mathematically that this should be so and, while this is not necessary here, the effect is so important that it is worth demonstrating how it comes about.

Suppose the duration of an activity is expected to average 30 days, but is subject to these disturbing influences:

— weather, distributed as in Fig. 6.2(a), with a tendency to cause delays;
— staff illness and absenteeism, which has a two-peaked distribution, reflecting a tendency for staff to be either generally fit, or all suffering from common ailments (Fig. 6.2(b));

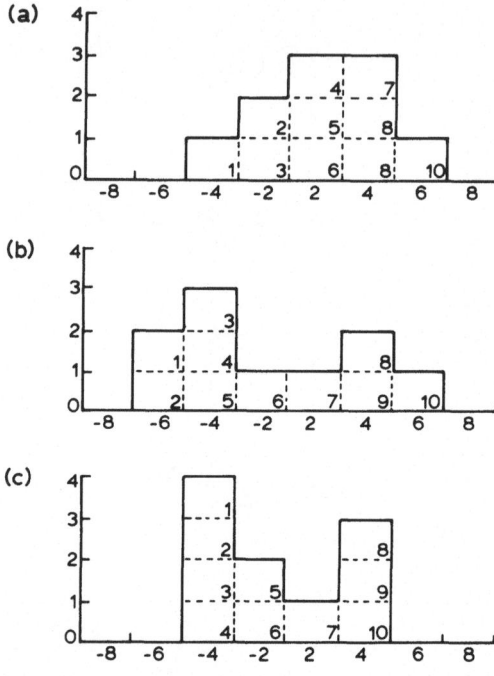

Fig. 6.2. Distribution for simulation example. (a) Weather; (b) absenteeism; (c) material. Horizontal scale:days delay or acceleration. Vertical scale:frequency.

— materials variation, again two-peaked, reflecting, say, the variety of bush to be cleared, which tends to be either fairly light, or rather heavy, with medium densities being rarer (Fig. 6.2(c)).

The method behind this demonstration of how such very different probability distributions combine to generate a close approximation to the normal curve will be dealt with in some detail, because it should provide an understanding of the process of simulation, which will be referred to again later. For each of the three probability distributions, the histogram has been divided into numbered rectangles of equal area, representing outcomes of equal probability. The choice of the number of rectangles as 10 is arbitrary. If three numbers in the range 1–10 are picked at random, and each is used to select a rectangle from one of the three probability distributions, we will have a simulation of one run of the activity, under the random influences of the three factors. Table 6.1

Table 6.1. Simulation of 20 Trials of an Activity, to Generate Probability
Distribution of Its Duration

Basic duration	Randomly chosen effect			Total of random effects	Total duration
	Weather	Absenteeism	Materials variation		
(a)	(b)	(c)	(d)	$(E = b + c + d)$	$(F = a + c)$
30	−2	+4	+2	+4	34
30	−2	+6	−4	0	30
30	−2	+4	−4	−2	28
30	−2	−2	+4	0	30
30	+4	−2	+4	+6	36
30	−2	+6	−4	0	0
30	+6	+4	+4	+14	44
30	+2	−4	−4	−6	24
30	+2	+4	−4	+2	32
30	+2	−2	+2	+2	32
30	+6	+4	−4	+6	36
30	+2	−2	−2	−2	28
30	+2	−4	−2	−4	26
30	+2	−4	+2	0	30
30	+4	+4	−4	+4	34
30	+2	−4	−2	−4	26
30	−2	+6	−4	0	30
30	+4	−4	+2	+2	32
30	+6	−4	−4	−2	28
30	−4	−6	+2	−8	22

+ Indicates a delay, − a reduction in time.

shows in detail how this was done for 20 trials, with the three influences
being added to, or deducted from, the average duration.

The word 'random' requires a little definition: in everyday use, it
means, taken haphazardly, whichever comes first to hand. This is not
satisfactory for this sort of application: if you asked people to give
numbers at random, you would observe biases, such as that many people
seem to 'like' even numbers, or the number three. In statistical use,
'random' means that every element of the population being sampled has
an equal chance of being selected at each trial, i.e. regardless of what
happened in the previous trial, or to its neighbours. A suitable choice of

random numbers could be made by drawing from a carefully shuffled pack of numbered cards; electronic devices known to generate random signals can be used; or there are mathematical procedures (called random number generators) that will provide a suitable series of numbers— these are the basis for most published random number tables, like those used to compile the example.

This family of simulation methods are often called 'Monte Carlo' methods, because of the obvious connection with gambling; they will crop up again later in this chapter.

Returning to this particular application of these methods— demonstrating how it comes about that a bell-shaped probability distribution can arise by the interaction of small random disturbances, each with almost any sort of probability distribution—we can see that the probability distribution of the sum of the average durations and the disturbances is, in fact, quite a reasonable approach to the promised shape (Fig. 6.3). It is a little skewed—i.e. there is a longer tail on one side than on the other—but otherwise the resemblance is very marked. How has this come about? The thin 'rim' of the bell results from the fact that only a very few combinations can produce the extreme values, e.g. a value as high as $30+16=46$ days can only be produced by the most extreme value of weather ($P_{(+6)}=0\cdot1$), of staff absenteeism ($P_{(+6)}=0\cdot1$), and of materials ($P_{(+4)}=0\cdot3$). We have already seen how to calculate the probability of these items all happening, by multiplying together their

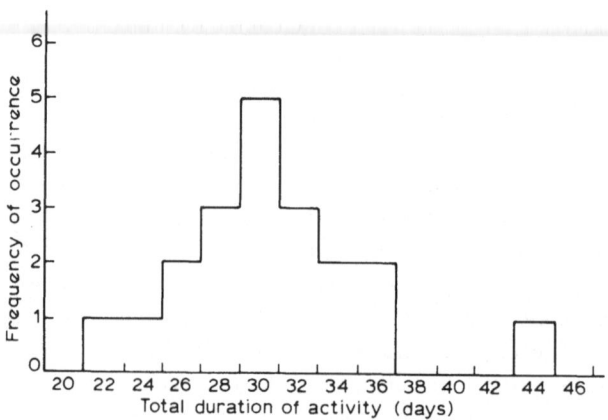

Fig. 6.3.   Frequency distribution of combined effects.

individual probabilities $P_{(+16)} = 0.1 \times 0.1 \times 0.3 = 0.003$, i.e. such an extreme value could be expected in 3 cases in 1000.

The flat summit results from the fact that there will always be a relatively large number of combinations around the mean that yield similar values, as may be verified by working out the number of combinations that give +2, 0, and −2 days, for example. Finally, of necessity there is a zone of steep fall-off connecting these two.

It has been proposed that various modifications of the normal curve should be used for activity durations: the normal curve extends to infinity in either direction, whereas durations cannot take an infinitely short time. (They could take an infinitely long time, if the project is cancelled, of course.) The modifications affect the tails of the distribution, which are of least interest; a commonly used variation is the so-called β-curve.

6.2.2. Statistical Processes
There are three main groups of processes which we will now consider.

(a) Digestion
If one has, say, 430 different activity times for past trials of a particular activity, before the information can be used, it has to be condensed into an idea, a concept, simple enough to manipulate. This could be done by averaging, but there is a choice of methods for even this most elementary type of statistical digestion. Other statistical digestion processes include the extraction of a trend (regression), and the estimation of how many real, distinct entities are being observed when several associated variables are being measured (factor analysis).

(b) Testing
Often, there is some idea, or hypothesis, either suggested by the data, or which was the reason for their collection, and it is necessary to test whether the observed effects are 'real' or chance outcomes of stochastic variation. The classic application is in field experimentation, e.g. to determine whether a particular husbandry practice affects crop yield; a more immediate example is detecting whether the accumulated experience of actual project durations warrants an across-the-board revision of the estimates.

(c) Prediction
Most prediction (outside of astrology) relies on discovering the factors

affecting the situation and assuming that they will continue to act in the same way in the future. Often, this is done on a very simple model, e.g. that the average work rate for a particular activity in the past will hold good in the future.

6.2.3. Chief Characteristics of Variables

Although there are other characteristics of variables that may be measured, the two we have discussed so far are the 'typical value' or, more accurately, central tendency, and of course, variation.

(*a*) *Central Tendency*
This tries to estimate what the 'real' value of the variable is. There are three measures in common use (there are others, applicable in special instances). These measures are shown diagrammatically in Fig. 6.4, and are described as follows:

— the median, i.e. the value which has 50% of the observations above it, and 50% below it; this measure is often of interest when the discussion concerns fairness, e.g. of income distribution;
— the mode, i.e. the commonest single value (there may be more than one mode, as in the case of the distribution of staff absenteeism effects in Fig. 6.2); this may be the most useful measure where there is a choice of actions, each appropriate to only one value of the variable being observed;
— the average (mean). This coincides with the other two only in the case of symmetric distributions, and has an important mathematical property: the mean of a sum is the sum of the means. (This is not necessarily true of either of the others.) This property means that, for example, the sum of the average durations of activities on a segment of a network is the average duration of the whole segment.

(*b*) *Variation*
It is worthwhile to look at the ways in which variation might be measured, to understand the reasons for—and the advantages of—the apparently clumsy measure used.

(i) The most obvious and tempting one, is the average deviation; a little thought will show that this is useless, as it will always be zero, with the positive and negative deviations cancelling out (if they do not, you have miscalculated the average!).

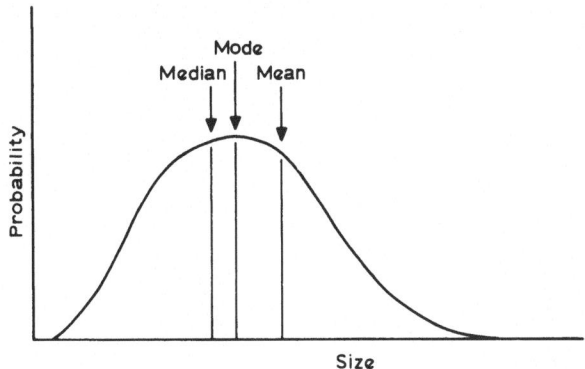

Fig. 6.4. Mean, median and mode for non-symmetric distribution.

(ii) This cancelling-out effect can be sidestepped by taking the mean absolute deviation (MAD), i.e. the mean deviation, obtained by treating all deviations as positive, and averaging them. This is quite a sensible measure, but has two drawbacks: the MAD of a sum is not the sum of the MADs; and MAD gives equal weight to a number of small deviations and a few large ones. The latter case is often more serious, in practical terms: if an activity duration has small variations, that may not be as serious, as the occurrence in, say, 15% of cases of relatively large ones. Table 6.2 shows two distributions, with the same MAD, but one series has the variation concentrated into fewer, larger, and potentially more disruptive deviations.

(iii) The next approach, suggested by the need to remove the cancelling-out effect of positive and negative deviations, is to square them, sum the results, and average them. In fact, this is the method used, except that the divisor is $(N-1)$, where N is the number of observations: there are sound mathematical reasons for using $(N-1)$, rather than N, but the intuitive explanation is that we are measuring variation, and that two observations only give one estimate of this, three give two, and so on. This measure of variation is called variance; it has ideal properties, being additive (so that, for example, the variance of the duration of a segment of the network is the sum of the variances of the component activities), and giving more weight to large single variations than to the same amount of absolute deviation distributed over several observations (the variances of series A and B, Table 6.2, are 2 and 2·9 units respectively).

Table 6.2. Comparison of Distribution of Two Series of Durations with the Same Mean Absolute Deviation (MAD)

	Series A			Series B	
Observed value	Average	Absolute deviation	Observed value	Average	Absolute deviation
12	10	2	9	10	1
9	10	1	13	10	3
10	10	0	8	10	2
11	10	1	10	10	0
12	10	2	10	10	0
11	10	1	9	10	1
8	10	2	13	10	3
9	10	1	9	10	1
9	10	1	9	10	1
9	10	1	10	10	0

Total absolute deviation = 12   Total absolute deviation = 12
Mean absolute deviation = 1·2   Mean absolute deviation = 1·2

The only problem with variance is the units: the squaring process affects these as well, so that the variance of a length is in square feet, and so on. Variances are, therefore, square-rooted, to get a quantity known as the standard deviation (SD); Table 6.3 shows, for normal variables, the distribution of the range, in terms of the SD.

Table 6.3. Relationship Between SD and Range for Normal Variables

Range, either side of mean, in units of 1 SD (a)	Approx. percentage of population included in limits under (a) (b)
0·67	50
1	70
2	90
3	99·7

6.3. EXPECTED DURATION OF CRITICAL PATH OR CRITICAL SEGMENTS

The traditional way of applying probability concepts to CPA has been to estimate the probability that events of interest (including final completion) will be achieved within some specified time, or to state the range of times likely for the completion of any particular activity at some stated level of probability.

From the discussion in the last section, it will be apparent that doing this requires an estimate of the variance for each activity that must precede the one under consideration—in all activities, if it is project completion that is being considered. Almost inevitably variance estimates will be based on subjective assessments of the 'likely' range; the general form of the relationship between range and SD has been set out in Table 6.3.

6.3.1. Probability Distribution of Duration

The method usually adopted is as follows: the section head is asked to state the most likely duration of each activity (this means the mode, not the mean); the fastest time in which he would expect the activity to be completed, in all but 5% of cases; and the slowest time he would expect, in all but 5% of cases. The mean time is estimated as

$$\frac{(\text{fastest} + \text{slowest} + 4 \times \text{likeliest})}{6}$$

and the variance as

$$\left(\frac{(\text{slowest} - \text{fastest})}{3 \cdot 2}\right)^2$$

The factors '4' and '3·2' arise from the precise shape of the modified (β) version of the normal distribution, and the squaring process converts the estimate of SD given by (slowest − fastest)/3·2 into the variance. (Even up to this point, very considerable demands have been made on the imagination of a man who is almost certainly statistically illiterate, for information that he may not have—if you think it is easy, try to give the fastest, likeliest and slowest times for a few of your routine daily tasks.)

Table 6.4 gives this information for a hypothetical project, of seven activities: a start-up activity, A (critical), from which branch a further critical segment (E and F) and a non-critical segment (B, C and D) which both converge onto a tidy-up activity G (also critical). These are shown in bar-chart form in Fig. 6.5.

Table 6.4. Estimating Variance and Mean Duration of Activities for a Hypothetical Project

Activity	Estimated			Mean	Variance
	Fastest (a)	Likeliest (b)	Slowest (c)	$\dfrac{(a+4b+c)}{6}$	$\left(\dfrac{(c-a)}{3\cdot2}\right)^2$
A	8	10	14	10·3	3·5
B	9	13	17	13	6·3
C	35	40	45	40	9·8
D	11	18	25	18	19·1
E	25	28	30	27·8	2·4
F	40	45	50	45·0	9·8
G	6	9	13	9·2	4·8

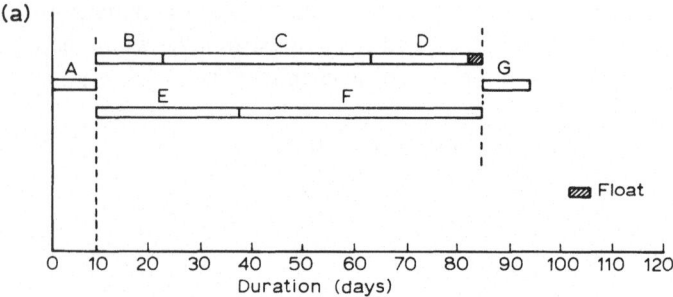

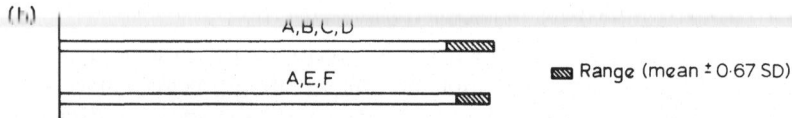

Fig. 6.5. Bar-chart of example: stochastic variation of duration. (a) Bar-chart— mean times. (b) Variation in duration of paths A, B, C, D and A, E, F.

In this method, the constant activity durations used before are replaced by the estimated means, and the forward and backward passes are made in the normal way. Because of the additive property of variances, these can also be added along segments, in just the same way: thus the variance of the duration of the critical path is the sum of the variances for A, E, F and G, i.e. 20·5. This corresponds to an SD of 4·5 days, on a mean of 92 days; and even the elementary information on the normal distribution already presented suggests we would expect the actual duration to be within ± 0.67 SD, i.e. ± 3 days of the mean, since this range should span 50% of the variation.

We can actually do quite a bit better than that: for the normal curve, the statistic Z, defined as (observation $-$ mean)/SD, is available in standard tables—one version is reproduced in Table 6.5. This can be read in two ways: if we want to know how likely it is that stochastic variation will carry the project more than 10 days over, we calculate Z as $(102 - 92)/4.5 = 0.014$, find the nearest value to this in the row and column headings part of the table, and look up the corresponding probability in the body of the table. (Note the difference from Table 6.3: that table refers to deviations either way, underrunning or overrunning; this version is for overruns only.) Alternatively, we can pick a probability, say 0.1 (10%), look up how large a value of Z this allows, and work back to the maximum size of overrun which has a '10% chance of happening':

$$Z = 1.28$$

$$1.28 = \frac{\text{extended time} - 92}{4.5} = \frac{\text{overrun}}{4.5}$$

Therefore overrun $= 6$ days (rounded up).

Deciding on either the level of probability to use, in the latter case, or the significance of the level of probability measured in the former, is not a straightforward matter; 0.10 is commonly used in both cases, but this is discussed further below.

The above procedure may give useful indications of the likely range of completion dates—but little more, given that the basis of the whole edifice of calculation is so shaky. It may indicate the need for crashing, to ensure that completion is reasonably certain to occur before some externally imposed date (e.g. beginning of a monsoon season)—but the trade-off between reducing uncertainty and increasing cost must be borne in mind, as must the possibility that any introduction of novel methods, to bring more resources to bear and thus speed up the work, may itself carry a penalty of increased variance, precisely because it is new and therefore likely to be less reliable.

6.3.2. Difficulties in Applying Probability Concepts to CPA

(a) The most basic problem is that the technical specialists may have

Table 6.5. Probability Distribution of Z

Second decimal place of Z_0

Z_0	0·00	0·01	0·02	0·03	0·04	0·05	0·06	0·07	0·08	0·09
0·0	0·5000	0·4960	0·4920	0·4880	0·4840	0·4801	0·4761	0·4721	0·4681	0·4641
0·1	0·4602	0·4562	0·4522	0·4483	0·4443	0·4404	0·4364	0·4325	0·4286	0·4247
0·2	0·4207	0·4168	0·4129	0·4090	0·4052	0·4013	0·3974	0·3936	0·3897	0·3859
0·3	0·3821	0·3783	0·3745	0·3707	0·3669	0·3632	0·3594	0·3557	0·3520	0·3483
0·4	0·3446	0·3409	0·3372	0·3336	0·3300	0·3264	0·3228	0·3192	0·3156	0·3121
0·5	0·3085	0·3050	0·3015	0·2981	0·2946	0·2912	0·2877	0·2843	0·2810	0·2776
0·6	0·2743	0·2709	0·2676	0·2643	0·2611	0·2578	0·2546	0·2514	0·2483	0·2451
0·7	0·2420	0·2389	0·2358	0·2327	0·2296	0·2266	0·2236	0·2206	0·2177	0·2148
0·8	0·2119	0·2090	0·2061	0·2033	0·2005	0·1977	0·1949	0·1922	0·1894	0·1867
0·9	0·1841	0·1814	0·1788	0·1762	0·1736	0·1711	0·1685	0·1660	0·1635	0·1611
1·0	0·1587	0·1562	0·1539	0·1515	0·1492	0·1469	0·1446	0·1423	0·1401	0·1379
1·1	0·1357	0·1335	0·1314	0·1292	0·1271	0·1251	0·1230	0·1210	0·1190	0·1170
1·2	0·1151	0·1131	0·1112	0·1093	0·1075	0·1056	0·1038	0·1020	0·1003	0·0985
1·3	0·0968	0·0951	0·0934	0·0918	0·0901	0·0885	0·0869	0·0853	0·0838	0·0823
1·4	0·0808	0·0793	0·0778	0·0764	0·0749	0·0735	0·0722	0·0708	0·0694	0·0681
1·5	0·0668	0·0655	0·0643	0·0630	0·0618	0·0606	0·0594	0·0582	0·0571	0·0559
1·6	0·0548	0·0537	0·0526	0·0516	0·0505	0·0495	0·0485	0·0475	0·0465	0·0455
1·7	0·0446	0·0436	0·0427	0·0418	0·0409	0·0401	0·0392	0·0384	0·0375	0·0367
1·8	0·0359	0·0352	0·0344	0·0336	0·0329	0·0322	0·0314	0·0307	0·0301	0·0294
1·9	0·0287	0·0281	0·0274	0·0268	0·0262	0·0256	0·0250	0·0244	0·0239	0·0233
2·0	0·0228	0·0222	0·0217	0·0212	0·0207	0·0202	0·0197	0·0192	0·0188	0·0183
2·1	0·0179	0·0174	0·0170	0·0166	0·0162	0·0158	0·0154	0·0150	0·0146	0·0143
2·2	0·0139	0·0136	0·0132	0·0129	0·0125	0·0122	0·0119	0·0116	0·0113	0·0110
2·3	0·0107	0·0104	0·0102	0·0099	0·0096	0·0094	0·0091	0·0089	0·0087	0·0084
2·4	0·0082	0·0080	0·0078	0·0075	0·0073	0·0071	0·0069	0·0068	0·0066	0·0064
2·5	0·0062	0·0060	0·0059	0·0057	0·0055	0·0054	0·0052	0·0051	0·0049	0·0048
2·6	0·0047	0·0045	0·0044	0·0043	0·0041	0·0040	0·0039	0·0038	0·0037	0·0036
2·7	0·0035	0·0034	0·0033	0·0032	0·0031	0·0030	0·0029	0·0028	0·0027	0·0026
2·8	0·0026	0·0025	0·0024	0·0023	0·0023	0·0022	0·0021	0·0021	0·0020	0·0019
2·9	0·0019	0·0018	0·0017	0·0017	0·0016	0·0016	0·0015	0·0015	0·0014	0·0014
3·0	0·00135									
3·5	0·000 233									
4·0	0·000 0317									
4·5	0·000 003 40									
5·0	0·000 000 287									

Reproduced from ref. 2, p. 590, by kind permission of John Wiley & Sons, New York, and R. E. Walpole.

neither the breadth of technical experience, nor sufficient feel for probability concepts, to respond usefully to the basic questions. In this case,
any attempt to force out some figures—any figures—to plug into the
method, can easily result in dangerously misleading results; it may also
weaken the respect of the technical specialist for the analyst, if the former
cannot see the point of the additional queries, and thus spoil what might
have been achieved by a straightforward non-probabilistic approach.

(b) The method ignores the possibility that, if the variance of the
duration of a sub-critical segment, e.g. BCD in our example (Fig. 6.5 and
Table 6.4) is great enough, stochastic variation alone will have a significant chance of ensuring that, during implementation, that segment
actually takes long enough to become critical. There is a rule of thumb
that suggests that, at any convergence, if the difference in total duration
of the two longest direct paths to that point is less than the larger of their
standard deviations, no correction for this possibility will be needed.[1]

However, in the example, there clearly is a problem: the SD values of
the durations of the paths AEF and ABCD are 4·0 and 6·2; the difference
of their durations is two days only; 50% of occurrences in the path AEF
would lie within the range 80–86 days (calculated as mean $\pm 0·67 \times$ SD),
of ABCD, within 77–85 days (calculated as mean $\pm 0·67 \times$ SD), so a shift
of the critical path, at implementation, is quite likely. There is no
solution to this problem—at least none that is usable in practice. By use
of the Monte Carlo simulation methods described above, it is possible to
simulate several repeats of the project, differing only in the size of the
effects of the stochastic disturbances. If enough of these are performed,
activities can be given 'criticality indices'; these indices measure the
probability that any particular activity will actually lie in the critical
path. Although this will indicate which areas of the project might benefit
by the provision of stand-by resources, the amount of computation will
virtually always exclude this approach, at least in our context.

(c) The traditional statistical approach ignores the complex effects of
stochastic variations in activity duration on the resource-demand profile:
as the length of each activity varies, so will the time-pattern of its
resource demands. In some cases, if an activity takes longer, the total
demand for associated resources will be extended proportionately; this
would apply to, say, bush clearing equipment; in others—especially
administrative activities—the amount of the relevant resource (professional staff time) may be unaffected.

(d) The procedure also assumes that the variances are independent, thus ignoring the fact that the project management system is—or should be—a reactive system. The assumption is wrong, because, if activities are overrunning, steps will be taken to accelerate later activities, so that there is some element of negative correlation between the durations of earlier and later activities; this means that variances can no longer be summed along segments of the network.

(e) Associated with the reactiveness of the project management system is the point that the object of the exercise is not to produce a perfect plan, but a sound plan and an efficient way of adapting that to the unfolding of events, because it is so unlikely that all the possible problems will have been foreseen. The things that will really throw a project out are the unforeseen problems, not the stochastic variations in well-recognised activities.

(f) The interpretation of the level of probability with which a certain completion date can be met, or the choice of the level of probability for the determination of a likely range of completion dates, present problems; this issue is discussed in Section 6.5, as it also applies to the next section.

6.4. TESTING FOR THE NEED TO REVISE DURATION ESTIMATES

Suppose that a project consisted of 100 activities; after 10 activities all in the same technical area and subject to similar disruptive influences—had been completed, it was observed that, on average, they were overrunning by 35%. We could say, here is a sample to 10 ratios; if nothing were amiss, we would expect them to have an average value of 1, apart from stochastic variations; because we would also expect them to have a normal distribution, we can calculate how likely it is that a value as extreme as 1·35 would have occurred, purely as a result of small random disturbances.

At first sight, the statistic Z looks useful—but that was applied to the large sample (of either real occurrences or mental reviews) of durations used in the estimation of variances; for small samples (say, less than 25) it is safer to use the statistic t (see Table 6.6), defined as:

$$\frac{\text{sample mean} - \text{hypothetical mean}}{\text{SD} \times \sqrt{n}}$$

Table 6.6. Probability Distribution of t

Pr df	0·25	0·10	0·05	0·025	0·010	0·005	0·001
1	1·000	3·078	6·314	12·706	31·821	63·657	318·31
2	0·816	1·886	2·920	4·303	6·965	9·925	22·326
3	0·765	1·638	2·353	3·182	4·541	5·841	10·213
4	0·741	1·538	2·132	2·776	3·747	4·604	7·173
5	0·727	1·476	2·015	2·571	3·365	4·032	5·893
6	0·718	1·440	1·943	2·447	3·143	3·707	5·208
7	0·711	1·415	1·895	2·365	2·998	3·499	4·785
8	0·706	1·397	1·860	2·306	2·896	3·355	4·501
9	0·703	1·383	1·833	2·262	2·821	3·250	4·297
10	0·700	1·372	1·812	2·228	2·764	3·169	4·144
11	0·697	1·363	1·796	2·201	2·718	3·106	4·025
12	0·695	1·356	1·782	2·179	2·681	3·055	3·930
13	0·694	1·350	1·771	2·160	2·650	3·012	3·852
14	0·692	1·345	1·761	2·145	2·624	2·977	3·787
15	0·691	1·341	1·753	2·131	2·602	2·947	3·733
16	0·690	1·337	1·746	2·120	2·583	2·921	3·686
17	0·689	1·333	1·740	2·110	2·567	2·898	3·646
18	0·688	1·330	1·734	2·101	2·552	2·878	3·610
19	0·688	1·328	1·729	2·093	2·539	2·861	3·579
20	0·687	1·325	1·725	2·086	2·528	2·845	3·552
21	0·686	1·323	1·721	2·080	2·518	2·831	3·527
22	0·686	1·321	1·717	2·074	2·508	2·819	3·505
23	0·685	1·319	1·714	2·069	2·500	2·807	3·485
24	0·685	1·318	1·711	2·064	2·492	2·797	3·467
25	0·684	1·316	1·708	2·060	2·485	2·787	3·450
26	0·684	1·315	1·706	2·056	2·479	2·779	3·435
27	0·684	1·314	1·703	2·052	2·473	2·771	3·421
28	0·683	1·313	1·701	2·048	2·467	2·763	3·408
29	0·683	1·311	1·699	2·045	2·462	2·756	3·396
30	0·683	1·310	1·697	2·042	2·457	2·750	3·385
40	0·681	1·303	1·684	2·021	2·423	2·704	3·307
60	0·679	1·296	1·671	2·000	2·390	2·660	3·232
120	0·677	1·289	1·658	1·980	2·358	2·617	3·160
∞	0·674	1·282	1·645	1·960	2·326	2·576	3·090

where n is the sample size. This is safer because the tables of t allow for the reduced information obtainable from a small sample.

If the SD of the durations in the sample is 0·2 units, t is

$$\frac{1·3-1}{0·2 \times \sqrt{10}} = 2·06$$

The t-table is entered with this value, on the row appropriate to the sample size; for this application, that is $n-1=9$ 'degrees of freedom'. (Degrees of freedom is a rather archaic term, and reflects the fact that, given a sample of, say 10, and a known mean, only nine of the individual values could be fixed freely and arbitrarily; tabulating the statistic this way makes the tables more versatile, and usable for applications in which the correct deduction is other than 1.)

We see that a value of this size would be expected to occur by chance with a probability of between 0·05 and 0·025 (actually, 0·038 by interpolation); we must accept that experience and the original estimates are not consistent, so that revision is essential.

This procedure does give a good feel for how unlikely it is that a particular overrun is due to chance alone; the cut-off point is often set at a probability level of 0·1, but this is arbitrary—a point discussed in the next section.

However, t-tables need to be used circumspectly: at each use, you compare t calculated for your sample of duration ratios, with the probability distribution of all possible cases of t, including those that might have been calculated in cases where the ratio was so close to 1 that the question of whether there was a consistent pattern of overruns would never have been raised. This means that, if you only use the test for extreme cases, you will bias the test in favour of giving the judgement that the overrun is a real deviation, i.e. an indication that something other than stochastic variation is at work.

The probability level must also be regarded properly: conventionally, 0·1 is used, i.e. ratios for which the t-value would only be exceeded with a probability of 0·1 or less are taken as indicative of a real deviation. (Such ratios are said to be significant at the 0·1, or 10%, level.) This does not mean that, in making such a judgement, you are 90% sure you are right; the probability relates to your pattern of action over several trials. If you adopt this pattern of action, accepting the t-test at the 10% level of significance, you will only make mistaken positive judgements in 10% of cases. (But you will probably never know which 10%—certainly not in time to do anything about it!) There are, of course, also mistaken negative judgements, where a real underlying drift is masked by stochastic disturbances so that a misleadingly small t-value is observed.

6.5. CHOICE OF PROBABILITY LEVELS

The problem of choosing an appropriate probability level has arisen in two contexts: the use of Z-tables for estimating the likely range of completion dates; and the use of t-tables for deciding whether observed activity durations are consistent with the original estimates. (The problem of interpreting the levels measured with these tests, in the alternative procedures of estimating the probability of attaining some specified completion date, or of getting as extreme a t-value for an overrun, is closely related, of course.)

The central issue is that the usual practice is to adopt some arbitrary value such as 0·1; this is objectionable, because it does not recognise the fact that the tolerable probability of a mistaken judgement depends on the cost of the mistake. This can readily be seen, if we compare two cases:

— a t-test of an overrun which, if positive, would imply a major reorganisation of the remaining phases of the project, with a cost increase of 12% in the total expenditure;
— a Z-test, carried out before implementation started, which, if positive, would imply crashing one lengthy, labour-intensive activity, at an additional cost of 1% of total expenditure.

The cost of a mistaken judgement will be much greater in the former case, and this should be reflected in the level of probability chosen; and a proportion of mistaken positive judgements—on average, equal to the probability level chosen—must be expected, since this is inherent in the nature of statistical tests.

There is a further problem: there can be mistaken negative judgements, where, say, an overrun is dismissed as the result of chance, when it is actually the outcome of a real difference in scale between estimate and actual; in this case, the cost is the damage incurred by failing to react. In general, the cost of the two types of errors—mistaken positive and negative judgements—will not be equal.

Statistical techniques—belonging to the family called 'Bayesian Methods'—are available to deal with these problems; readers who are interested in these will find a good account with minimal mathematical content, in Reference 2. However, these will involve further subjective probability estimates, and, in most cases, it is probably better to accept that the statistical methods described earlier improve the project manager's 'feel' for the situation, rather than to build an even taller structure of calculations on the same, already shaky, foundations.

6.6. DECISION NETWORKS

Often, particularly at early planning stages, there may be considerable uncertainty about the actual outcome of various activities of an exploratory nature. These will tend to alter the form of the network itself; the technique of decision networks provides a simple way of analysing the consequences.

Consider the case of a project to upgrade an extension service; a key part of this process was the provision of housing for field assistants, both to enhance their status, by demonstrating that they were valued servants of government, and to make them independent of the larger landlords, who had traditionally provided field assistants' housing—nominally free, but actually in return for various services, including part-time assistance in managing the landlords' farms. Three options were available:

— renting accommodation, which was almost certain to be cheapest, and least likely to result in housing provision becoming a critical activity delaying completion, but not always feasible;
— purchasing prefabricated housing, which would be quick to install, but would require scarce foreign exchange;
— building traditional houses, which would be slow, but low in foreign exchange costs.

The relevant parts of the network are shown in Fig. 6.6; the new element is the presence of the diamond-shaped nodes; there are two exit arrows from each (there could be more), each with the probability (usually estimated subjectively) that that particular outcome would result. (The probabilities on the arrows emerging from any one diamond add to one: they are alternatives, therefore the probability that one or other will occur is the sum of their individual probabilities, and it is certain that one or other will occur, so the total must be 1.) More elaborate schemes of representation are in use.[3]

Using the rule for calculating the probability of things happening together (Section 6.2.1), the following probabilities result.

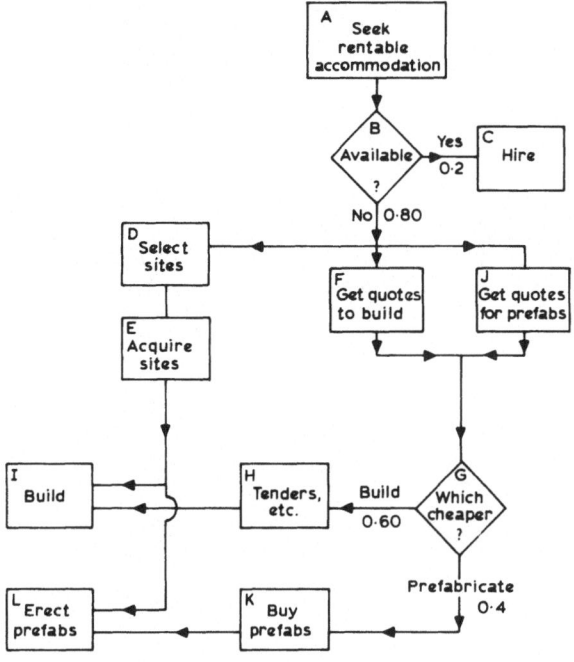

Fig. 6.6. Decision network.

— There is a probability of 0·2 that rented accommodation at any one site will be possible; in this case, this part of the network reduces to A, B, C.

— There is a probability of 0·8 × 0·6 that rental is not possible, and building will be cheaper, i.e. there is a probability of 0·48 that this part of the network will reduce to A, B, D, E, F, J, G, H and I.

— There is a probability of 0·8 × 0·4 that rental is not possible, and prefabrication will be cheaper, i.e. there is a probability of 0·32 that this part of the network will reduce to A, B, D, E, F, J, G, K and L.

If there was a large number of sites, these probabilities would reflect the proportions of the various outcomes expected, and would enable the management advantages of standardising on one option (other than rental) to be evaluated; if there was a single site (e.g. if the building work had concerned a major central building), the probabilities would relate to the network having one or other form, each with its associated completion time.

REFERENCES

1. Moder, J. J. and Phillips, C. R. (1970). *Project Management with CPM and PERT*, Chapter 11, Van Nostrand–Reinhold, New York.
2. Wonnacott, T. H. and Wonnacott, R. J. (1972). *Introductory Statistics for Business and Economics*, Chapter 15, Wiley, New York.
3. Moder, J. J. and Phillips, C. R. (1970). *Project Management with CPM and PERT*, Chapter 6, Van Nostrand–Reinhold, New York.

CHAPTER 7

Critical Path Methods at the Planning Stage

7.1. INTRODUCTION

We have seen that CPM does for the timing of project activities what budgeting does for their cost: it ensures that an allowance is made for all the items involved, in their proper relationship with each other, with no gaps, and no double counting. This means that, as far as is humanly possible, a feasible set of schedules for the timing of project activities is created; that is, feasible in terms both of the ways in which activities depend on each other, and in terms of the total resources required by all the activities expected to be in progress at any one time.

The great advantage of having schedules with this property is that the project manager and the client organisations for whom the work is being done have a set of attainable targets against which to measure physical and financial progress. There are few exercises which consume time less productively than holding a continuing inquest each month, or quarter, on why an undertaking is failing to reach its targets when those targets are physically unattainable. Progress forecasting becomes impossible, and, worse still, because the cause of the gap between planned and actual progress is likely to be wrongly identified if CPM are not used, the wrong remedies will be applied, generating waves of confusion that will spread out through the project.

We have discussed the main CP techniques available to stop this happening: production of the network, calculation of the initial timetable and critical path, adjusting this to take account of the resource constraints, and methods of compressing the schedule. All this was discussed as though the project was relatively simple in organisational structure, with no contractual problems, and was already underway, with a project manager and at least some senior staff in position. Often, in fact, CPM are introduced at just such a late stage, which usually leads to unnecessary problems. These are:

— that the basic framework of ideas about project duration is already firmly established, on an unrealistic basis, and this framework is difficult to modify, since it may be built in to agreements with donor agencies and so on;
— that the organisational structure of the project is such that no method of project control, CPM included, can be effective;
— that the contractual structure of the project is such that the project manager has insufficient control over contractors and sub-contractors to ensure that they organise their work in such a way that important activities which can delay other agencies' work are completed on time.

These problems are discussed in the following sections, with organisational structure problems presented in terms of the sponsoring agency and its sister organisations, and contractual problems, in the context of contracted and sub-contracted civil engineering works. This latter emphasis is because of the considerable importance of civil engineering works, in cost terms, on many development projects; in addition, civil engineering contracts deserve special consideration because, being very much one-off jobs, they are considerably more prone to generate problems than are contracts for most other goods and services required in development work. Section 7.6 concludes with a discussion of the application of CPM to the planning process itself.

7.2.   GETTING THE FRAMEWORK RIGHT

The framework—the overall phasing of the project in relation to policy makers' expectations, the availability of funds, commitments to other departments or ministries, and support given to, or required from, other projects—is most often unrealistic purely as the result of neglect. The difference in the number of professional man-hours usually devoted during planning to the economic and financial aspects of major projects and to the scheduling of project activities is enormous: in a number of major projects known to the writer the ratio between the two is wider than 50 to 1. There appears to be no real explanation for the relative difference in emphasis, apart from the fact that planners are more frequently trained in economics than in any other discipline. The difference is particularly difficult to understand on several counts.

First, it should be realised that an accurate estimate of the discounted rate of return on the project investment—the key measure of the project's potential economic worth—is itself dependent on the timing of expenditure relative to income: getting the schedule wrong at the planning stage, in such a way that delays to the inward flow of income are overlooked, could easily overestimate the attractiveness of the project as an investment. Secondly, the phasing of activities, and the overall duration of major groups of activities, can have an enormous effect on the price estimate that should be used: where construction is involved, for example, the price will depend very much on the period available for its completion, with shorter completion times normally attracting a higher bid. If the planner allots an unreasonably short period, and does his estimates properly, he will come up with an unnecessarily high price, and vice versa. Thirdly, traditional methods of scheduling underestimate duration, and therefore the inflation component. These are all points at which there is an internal inconsistency in the planning process, and are additional to the effects of failing to schedule properly on the smooth interlocking of activities, and on staff working habits and morale, which can impair the effectiveness of the project.

If no attempt is made to ensure that the time requirements of major groups of activities are properly budgeted for at the outset, the whole of the basic framework of the project will be distorted: the planners' estimates will be built into applications by the immediate promoters of the project for funds, sanctions to engage staff, and so on. If these are approved—usually with arbitrary cuts and changes to take account of any general shortage of resources—even if the overall amounts are correct, the phasing of the release of funds will often be wildly wrong. If the sanctioned pattern of funding is too slow, then it will delay project implementation; if it is too ambitious, there will be apparent underspending. The latter is by far the more common case; at first sight, this situation is relatively harmless, but, because allocations of funds are usually revised in the light of actual expenditure, it can be very damaging: the allocation of funds for a period following that in which an unexpected delay occurred may be cut, when the project management is already committed to spend them. Even in this situation, CPM can be of use: it is a lot easier to bargain with a financing department, against a cut in allocations, if one has a demonstrably sound analysis of what caused the under-expenditure, and what is being done to reduce the problems, rather than a limp promise that progress will somehow be better next time.

It is not only the financial aspect of the framework that is disturbed if the scheduling exercise is skimped by substituting an intuitively drawn bar-chart for a proper analysis of the project schedule: the client organisation's expectations of what is a reasonable rate of progress will have been formed around a schedule which is very likely to be impossible to implement, and the damaging process by which its project management is sent off—all unknowing—in pursuit of the unattainable, has been set in motion.

Right through the discussion of critical path techniques, we have seen the problems created by getting the basic framework of the project wrong: how, even at the end of the networking phase, project staff can be faced with the embarrassing realisation that the time allowed for completion is too short; how this can be made worse during the resource scheduling phase; the difficulties of a late attempt, with project staff already on the job, to introduce more resources to bring the completion date back to some pre-determined date; and the problems of an over-ambitious specification being adopted.

If the constraints imposed by the interdependence of project activities and their demands for resources are incorporated by proper planning at the outset, many of these problems will never arise; unreasonable expectations either of completion time (without corresponding measures to speed up work), or of the price of achieving some early completion date, will have been stifled at birth.

All this applies with equal force to a contractor, for whom the project means discharging his contractual obligations within his cost estimates: if there are unexpected delays, his bill for overheads—regular labour, office services, supervisory staff, finance charges and so on—will become disproportionately high. Failures of contractors as a result of their failing to plan the scheduling of their own work carefully are not uncommon, and their difficulties are compounded by the fact that most larger contractors will be relying on the flow of funds from projects in their most profitable phases to finance others which are in the relatively lean times of project start-up or wind-down.

Clearly then, it is essential that some arrangement is made for a preliminary CPM analysis of the project schedule; the next section deals with the inherent problems of getting the right data.

7.3. PROBLEMS OF DATA GATHERING AT THE PLANNING STAGE

The major difficulty is, of course, that the technical manager and section heads are not available; at the early planning stage, probably not even prospective contractors are available, and few of these would be prepared to give any significant assistance unless they were reasonably sure of at least being asked to tender.

However, there are potential sources of information:

(a) For administrative and organisational activities—procedures for obtaining release of funds, for tendering, hiring staff, and so on— the sponsoring organisation's resources will always be available.

(b) Activities carried out by sister organisations can also usually be investigated in considerable detail—sister organisations being associated firms in the case of a commercial organisation, or other government agencies in the case of a government department; this might include data on building, roadmaking, etc., from a Public Works Department (PWD).

(c) Information on general construction—housing, roadmaking, etc.— can also be obtained from a quantity surveyor, or a consultant engineer—but usually at a price, unless the client organisation has suitable staff of its own.

(d) Highly specialised sub-contractors for such items as milk processing plants, abattoirs, and so on, are more difficult. Equipment suppliers will often give information, if there is a reasonable prospect of at least being asked to tender; aid agencies may be able to provide a consultant (provided that arrangements have been made early enough); and, in some fields, commercial consultants may be available.

Between them, these sources should cover most items of information likely to be required—bearing in mind that aggregated activities will be used for CPA purposes at the planning stage. This aggregation has to represent a sensible compromise between the available time, and the risk of overlooking linkages between activities, and consequently underestimating project completion time. However, by far the most damaging omissions usually concern linkages between activities belonging to different disciplines, and these are relatively easy to identify.

One type of activity for which it will always be difficult to get any information is one in which some unusual method of construction or fabrication is being used, of which the concern responsible has no experience. By far the best way to handle this situation is by breaking the activity down into components whose durations are known, and building up the overall duration from these. This process may reveal that the proposed method is contributing an undue amount to total completion time, at a stage when it is still possible to adopt alternative, simpler methods of construction.

7.4. ORGANISATIONAL STRUCTURE OF THE PROJECT

The relevance of this is that the reporting system (Chapter 8) will only be as effective as the decision-making structure it feeds into. At the planning stage, this may still be flexible, and there may be scope for modifying it in ways which will improve the functioning of the project management system.

Our earlier surveys of the instruction and feedback links of the project management system assumed that there would be a sound chain of command. There are several pre-requisites for this.

(a) The first essential element is that there should actually be an identifiable individual project manager, a single point through which all decisions not taken at lower levels must pass, either for resolution, or, if there are serious disputes or major policy issues, for review and reference upwards to the appropriate authority. Without this, the channels for instructions and information can easily become duplicated, with the result that misunderstandings proliferate, and opportunities are created for playing both ends against the middle: disgruntled project staff who cannot get what they want through one channel, will try another. This sort of situation readily arises where a major project is shared between two government agencies, both reluctant to relinquish control of staff or funds to the other; the division of responsibility for agriculture, in the broad sense, into agriculture (meaning field crops) and livestock husbandry, and the division of responsibility between agriculture and irrigation, seen in some countries, tends to generate examples of this.

Often, the major responsibility for the project is given to the department handling the bulk of the funds, which means inevitably, the one responsible for any engineering works; this, 'senior', department may well have an individual described as a project manager. However, the 'junior' partner in this alliance may then provide staff—agricultural officers, agricultural engineers—at say, district level. These men may have other responsibilities outside the project; even if they do not, they will be part of a hierarchy of commands and funding for which the project is only one problem among many others. As a result, issues outside the project will at times drain money and man-hours away from it, and the 'senior' department's project manager will have no effective control over this. On paper, in this sort of situation, he can get remedial action taken by applying pressure through his department and through the upper levels of the other department, to the man he wants to influence. In practice, this just doesn't work, because it takes too long, and no one man bears responsibility for failure, or has the motivation to push things along.

The amount of administrative adjustment required to establish a unified structure may be enormous: it is well worth the effort required, to ensure that large sums of money are spent effectively, but unless the necessary steps are identified early enough, it will never be possible to achieve much. The problems often centre around organising the flow of funds where different agencies share a project: usually, the method by which the total sums available for capital and recurrent expenditure, and the persons who can disburse them, are defined in regulations which have the force of law. Often, it is simply not possible for a single ministry or department to choose to relinquish their statutory control, and formal arrangements may need to be made, at high level, to enable this to be done. Similar problems may occur with the administration of staff discipline and promotion.

(b) The second major requirement is that the project management team must have the necessary control over the agencies working on the project: this applies with particular force to contractors, and a later section suggests how it may be achieved in that case. The project manager must have control over certain key aspects of how such agencies operate—the order in which areas of land are cleared, or sections of an irrigation network are developed—where conflict and delay can arise between different disciplines. In the case of work contracted out, while the project manager must not attempt to meddle in the contractor's day-to-day running of the job, he must, by the terms of the contract, have control over these key items. Similarly, if the agency involved is a sister government body, such as a PWD, that body should not be free to organise its work—possibly to fit in with non-project obligations—at the expense of the project.

(c) The third main requirement is that the project manager should be—by training and experience—a manager rather than a technical expert. His job is to organise information and instructions so as to implement the project plan, not to act as the most senior technical expert on the project. Partly this is because someone has actually to manage, and it is unlikely that a very highly qualified technician will also be competent to do this. Partly, also, it is because a man selected for his technical expertise may tend to lose sight of the overall objective (which is to bring a productive agricultural project into being) either by getting side-tracked into interesting technical problems, or by an uneven appreciation of the problems of the different groups involved. In the fairly common situation of a civil engineer in charge of what is ultimately an exercise to grow more crops, the project manager may have more sympathy with the technical and financial problems of his opposite numbers in say, the civil engineering contractor's management, than with the intended beneficiaries of the project, the farmers in the area.

The writer has seen this situation develop on a large irrigation scheme: incompetent contractors were allowed, by a civil engineering oriented management, to destroy an existing irrigation network, almost completely, over substantial areas of a project, with dire effect on the livelihoods of numerous farmers. The mechanism of this was as follows: the project's sponsors had initially ignored the engineers' recommendations against the particular contractors and accepted their (very low) bid; the contractors had, in fact, bid so low that there was very little profit in the job, and had, in any case, inadequate financial resources. They therefore had a perennial cash shortage, and one of the easiest ways of coping with this was to drop other work, and to forge ahead with the excavation of the main drainage system on the scheme, which was easy work, and into which some of the profit element had been loaded (see Section 7.5.2). The project management team were forced into the choice of pressing the contractor to reinstate the old irrigation system (as he was contractually obliged to do), and thereby increase his troubles, or letting the farmers suffer. While there was a very real possibility that the extra pressure might have put the contractor out of business, it was also clear that the management had not even troubled to check over the extent of the problems that disruption of the irrigation network was causing to the farmers; these problems were serious enough to affect the long-term efficiency of the project.

(d) We have used the expression: 'chain of command'; this is commonly taken to imply that the authority structure of the administering agency is the traditional tree-type structure, seen in so many organigrams. There are alternatives (Fig. 7.1) and all should be considered, particularly in relation to the project sponsor's organisation, since this is the area in which feedback is turned around and converted into instructions: if this doesn't work well, neither will the rest of the system.

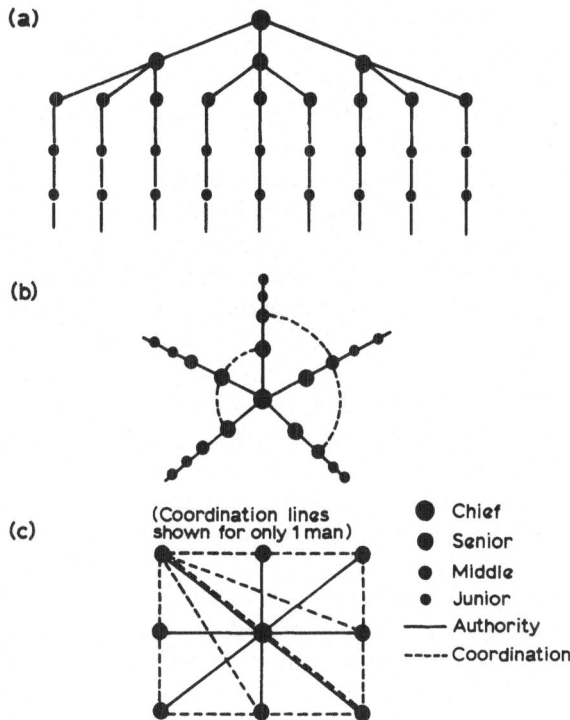

Fig. 7.1. Alternative organisation structures. (a) Greek temple; (b) power; (c) net.

(i) The traditional structure is sometimes referred to as a 'Greek temple' format (Fig. 7.1(a)), because of its shape. It is the ideal arrangement for managing steady, stable situations; roles, job descriptions, what each man is expected to contribute (not only under his formal contract, but also under his psychological contract with the organisation, in terms of unwritten requirements regarding conduct, out-of-hours work, professional self-improvement, etc.) are all clearly defined, as are the channels of communications; function is determined by job description, and there is no call for individual inspiration or conspicuous achievement, which may even be disruptive. It is, however, not the ideal arrangement for implementing a project: rigidity of procedures tends to lead to slow or irrelevant responses to unforeseen problems. It is still likely to be the best structure in cultures where respect for authority and seniority are important, because it is the only one that works well in those cases; it is also a very stable structure, in the sense that deficiencies in individuals tend to be compensated for by the very rigidity of the procedures inherent in such a formal system.

In practice, the growth of cross-linkages between the 'columns' of the 'temple' in complex, difficult, and changing circumstances, tends to convert this pattern into something more like format (iii), below, which is more appropriate. These linkages take the form of consultation groups, working parties on particular issues, and so on.

(ii) The wheel, or power, structure (Fig. 7.1(b)) is one in which all the main disciplines report directly to one man, at the centre, with some sideways coordination at certain levels of seniority; given a good manager this can be a very responsive arrangement, in that decisions and changes can be made promptly. However, there are corresponding weaknesses; any system which responds rapidly—whether it is mechanical or social—also tends to overshoot, or over-react, more than one which responds slowly; and there is no reinforcement in the form of procedures, to strengthen this type of structure if the man at the centre is not effective. Organisations in some Arab countries, drawn up on paper as 'Greek temple' in format, actually operate as wheel structures, sometimes to the confusion of outsiders: many people with a Western background do not even recognise the actual structure as being structured at all, because they are only aware of the type (i) format.

(iii) The net format (Fig. 7.1(c)) is one in which there is no permanent formal hierarchy, with the lead being taken on particular occasions by the person most relevant, and all channels of consultation being permitted. Obviously, this only works with staff of a high standard of professional commitment, and no cultural limitations on the assumption of responsibility by any individual, regardless of his age or status.

There is, of course, no reason why the same structural pattern should be adopted throughout: an appropriate mixture of power structure for central management, traditional hierarchical structure in costings and other technical departments, and net format for any division involved in trouble shooting, including CPM applications, could be appropriate. The important point is to realise that there is a choice, and it has to be an appropriate one, in the contexts of the task of the particular part of the organisation, its staff, and the culture in which all are embedded.

Handy[4] gives an excellent overview of current organisation theories, in the context of Western cultures.

This section would not be complete without some reference to the possibility of setting up a quasi-independent agency to implement a project: this gives greater freedom of action, at a cost in delay, and with the possibility that it may have to be disbanded at the end of the implementation phase.

7.5. CONTRACTUAL STRUCTURE OF PROJECTS

The relevance of this is that the contractual arrangements have to allow both the data gathering and instruction-issuing phases of the project management system to work effectively across the gap created by the fundamental need for the contractor to be able to act freely in discharging his obligations.

Any contractual arrangement has to recognise that, during implementation, the interests of the parties may be in conflict, at the interfaces between their various areas of responsibility: these are the points— convergences in the network—where the activities carried out by the various parties have to be controlled, to ensure that (preferably) the early start date for the next activity is met, or (at worst), its late start date is not passed. The conflict of interest arises because, for each party, the project consists only of his contractual obligations, and it may not pay him to go beyond these in adapting his programme to that of the others. Such adaptations may actually cost him money, in the obvious sense of extra expenditure on, say, extra labour, or extra overtime; it may also cost him money, in the sense of delaying his cash inflows, resulting in increased costs for financing the job. Similarly, the interests of the project's sponsors and those of the contractor, or contractors, are not the same; there is obviously a certain amount of common interest in keeping the project moving, but ultimately, the contractor is there to extract a profit from the sponsor, who, if he is sensible, is out to keep that profit to the minimum consistent with getting the job done satisfactorily. If the commercial relationship between the sponsor and contractor includes the possibility of further contracts in the future, as an inducement to the contractor to be cooperative, and if there is a shortage of suitable contractors, as an inducement to the sponsor (promoter is the standard civil engineering term) to treat the contractor considerately, some reliance may be placed on goodwill to ensure cooperation. If, at the other end of the spectrum, there is little likelihood of future work, or the competition between suitable contractors is intense, even the most scrupulous contractor will be unwilling to incur significant additional expense to resolve what really is someone else's problem. That means, it has to be made to be his problem, by writing appropriate conditions into the contract. In either case, once things start to go wrong, goodwill tends to evaporate rather rapidly!

What can be done about all this? Well, the planners can do the following:

— ensure that reasonable duration estimates are used for all the activities from the outset, and that a sufficiently complete inventory of activities is compiled to ensure that the interfaces in the network are all identified, and the critical areas highlighted, so that contractors are presented with a reasonable schedule;
— put pressure on the project's sponsors (promoters) to recognise that the assumptions built in to the preliminary network regarding the level of competence and technical accomplishment required imply a need for careful selection of potential contractors;
— as a corollary to this, where the promoter has adopted a policy of using local contractors (or contractors from other Third World countries), if this means using less competent or less experienced parties, ensure that this is reflected in the time and resources framework of the project;
— ensure that any essential restrictions relating to the scheduling of interfacing activities are properly written into the contracts;
— ensure that the contractual arrangements do actually allow the project management system to get a grip on the situation, in terms of extracting information for adjustment of schedules, and promoting a reasonable amount of compliance with the advice generated.

The last point is especially relevant: it is not so useful to write a set of rigid restrictions regarding the scheduling of interface activities, when, as we have seen, the critical path may actually snake about inside the network, under the influence of uncontrollable variation in activity durations (not to mention incompetence, omission of activities from the plan, and so on). The compliance mentioned is often more a matter of incentives than compulsion: either avoiding inadvertently generating incentives not to comply with the project management system, or providing positive incentives to cooperate.

To see how these five activities may be carried out, it is necessary to examine the ways in which contracts are drawn up, placed, and executed. These are fairly closely controlled in almost all countries; throughout the English speaking part of the Third World, they tend to follow the British system, which is defined in the code of practice set up by the Institute of Civil Engineers.[1] Similar codes are produced in other countries, and for other engineering disciplines. A summary of the normal practices is given in the next section, as a prelude to detailed discussion on how the above five desirable activities can be carried out; and Section 7.5.2 looks at relevant aspects of contractors' responses to the codes of practice.

7.5.1. Civil Engineering Contractual Practice

(a) Parties to the Contract and Their Initial Responsibilities
There are usually three parties involved: the Promoter (corresponding to our sponsor), the Contractor (or contractors) and the Engineer (either an independent professional, engaged by the Promoter, to protect his interests, subject to safeguards for the just treatment of the Contractor, or a salaried employee of the Promoter).

The engineer usually enters the picture at an early stage of the project, and his judgement on any major technical possibilities or problems of the site may well help to mould the form of the project, by adding or deleting components before the detailed planning stage. The client's consulting engineer—usually a firm rather than an individual—takes responsibility for the site investigations, which on an irrigation project, would include agricultural soils investigations, since the result of these obviously has a very profound impact on the scope and layout of the project, right down to the design of canals (since their capacities will depend on the crop rotations chosen). The engineer is also responsible for detailed design work prior to tender, advising on the type of contract, and preparation of contract documents. Ideally, fully detailed drawings would be provided as part of the contract documents, but often this is not possible; again the large irrigation scheme provides a good example, where it is normally impossible to complete the necessary topographic survey work before construction starts.

(b) *Types of Contract*
Possible types of contract include:

(i) Cost reimbursement contracts, in which the contractor is paid his actual costs, plus a fee, which may be a fixed sum, a fixed percentage of the costs, or an amount calculated on a sliding scale.

(ii) Target contracts, which are a variation on cost reimbursement contracts in which a target cost is agreed, and the contractor's fee is increased or reduced, according to whether he keeps costs below, or allows them to rise above, the target (the intention being to eliminate the obvious objection to straight cost reimbursement, which is that it offers a disincentive to the contractor to be economical).

(iii) Measurement contracts, which may be classified as:

— schedule of rates contracts, in which the contractor quotes rates for various types of work, to be executed anywhere in a specified area, the rates commonly being valid for a period of one year;
— Bill of Quantities contracts, in which the contractor puts prices on a detailed list of works, to each of which is attached an appropriate quantity, the final payment being the product of the quoted rate and the actual amount of work done, as measured during implementation.

(iv) Lump sum contracts, in which the contractor is responsible for his own estimates of the types and quantities of work to be done, and undertakes to complete the job for a set amount.

Of these, cost reimbursement contracts, with their obvious drawbacks, are probably most often used for urgent work (when they may be negotiated directly with a single reliable contractor); they are also most appropriate to industrial research and development work, where the actual scope of the job is not known at the outset. Schedule of rates contracts, offered for competitive tender, are often used for scattered items of work, such as canal cleaning, remedial roadworks, and so on. (Leading rates contracts are a variation on schedule of rates contracts, in which rates are agreed for only the most important items, in the interests of getting work started quickly.) However, for the civil engineering components of most agricultural development projects, the Bill of

Quantities contract, let by competitive tendering, is usually adopted, as being the safest and fairest procedure for all parties. All-in, or turnkey contracts, in which the promoter states his requirements in broad terms, and asks potential contractors to submit designs and quotations, are also used for large agro-industrial projects.

(c) Pre-qualification of Contractors

To save time and cost in scutinising a large number of competing bids, some of which may be unrealistic, some form of pre-qualification process is often adopted. This should eliminate bidders whose resources or experience are inadequate; the engineer plays—or should play—an important role in this, and other planners should reinforce this by making clear to the promoter the possible consequences of selection of an inadequate contractor.

(d) Contract Documents

Selected contractors are provided with a set of contract documents, which normally consist of:

(i) Instructions to tenderers (date and form of submission of tender, admissability of additional alternative proposals for parts of the work, arrangements for site visits, and so on).

(ii) Conditions of contract (the general terms of the relationship between the promoter and contractor, with their respective rights and obligations). Various standard sets of conditions of contract exist.

(iii) The specifications (detailed descriptions of the style, materials, and workmanship required for each part of the works).

(iv) The Bill of Quantities (which will cross-refer to the specifications).

(v) Drawings.

(vi) The Form of Agreement (the legal document which, when properly signed by the contractor and promoter, binds the contractor to execute the works, in accordance with the rest of the contract documents).

Where restrictions are required on the methods or completion dates for parts of the work, these are inserted in the specification. It is important to realise that these must represent the essential minimum of interference in the contractor's freedom of action. The contractor completes the Bill of Quantities, which includes the prices of not only material and direct labour, but of all associated costs (inspection, storage, wastage, temporary works such as temporary bridges and security fencing, cleaning up the site, and so on); these include the project's share of head office overheads, financing charges and profit, which may have to be shown separately.

The contract may include a formula for adjusting prices in the light of inflation, and may make provision for advances to the contractor for the purchase of specialist plant.

When he returns the contract documents—often at the end of a series of last-minute consultations with sub-contractors—he may be required to execute a bid bond: that is, he provides security that, if his bid is accepted, he will actually sign the contract. This may take the form of depositing a certified cheque with the promoters.

(e) Scrutiny

The bids are checked for errors and signs of misunderstandings; these include errors made by the contractor against his own interest, since, although it may be true that, in law, and in most countries, when the contract has been signed, the parties must perform their respective obligations, there is no way of preventing the contractor going into liquidation if he gets into serious difficulties. (This may have worse effects on the promoters than on the principals of the contracting firm.) Rates are compared between contractors, to see if any are unrealistic, or if there are signs that some bids are unbalanced (see Section 7.5.2). Finally, the engineer makes a recommendation to the promoter on which bid should be accepted; this may not be the lowest bid, if there are reasons to think that the bidder cannot actually perform what he offers at that price.

(f) Acceptance

After any errors and misunderstandings have been clarified, the promoter gives the contractor a letter of acceptance (which usually stipulates that future correspondence is to be directed to the engineer); the Form of Agreement is completed, the engineer formally sets a starting date. If required, a performance bond—providing some protection against default by the contractor—is also completed; amounts vary from 10% of the tender sum upwards, and in some countries, a separate bond to guarantee payment of workers is also required. (The promoter has further security, in that he will normally hold the retention fund—see below—and can usually seize the contractor's plant if the latter defaults.)

Tenders are generally published, and this information can be used by contractors in formulating future bidding policy.[3]

(*g*) *Execution*

The job is now underway; the engineer usually appoints a Resident Engineer (RE), who is responsible for issuing instructions and drawings to the contractor, and supervision of testing, measurement, workmanship, and of working methods; he also records progress, and is responsible for maintaining records (usually in the form of 'as constructed' drawings) of what is actually constructed. His counterpart on the contractor's staff is the Agent; the Resident Engineer and the Agent between them settle matters of fact when there is any claim or dispute regarding whether particular items of work and expenditure are within the scope of the contract. Variations on the contract are formally notified by the Resident Engineer to the Agent. It is desirable that these are kept to the minimum possible number by completing all design and investigation as early as possible: variations are a potent source of conflict, with either the contractor believing he has the promoter 'over a barrel', or that he is being unreasonably imposed on.

Sub-contracts require the engineer's approval; there are also nominated sub-contractors, concerns which the promoter has decided are specially suited to carry out parts of the work, by reason of the quality or appropriateness of their products and services (e.g. a supplier of modular canal outlets might be appointed on this basis). The contractor charges an agreed percentage on their charges. Technically, he has a right to refuse to accept a particular nominated sub-contractor (since he would be liable for the latter's default), but this may not be realistic if the subcontractor's work is in any way basic to the design of the main part of the works.

During execution, measurements of the work completed are made and certified to the promoter, who forwards payments to the contractor; usually a percentage of the payments is withheld by the promoter as a retention fund. Progress is recorded, and the contractor may become liable for liquidated damages: these are penalties for late completion, which in most countries cannot be enforced by the courts if they are more than a fair reflection of the actual losses incurred by the promoter as a result of the delay. An additional point is that, in at least some countries, it is also a valid defence against the imposition of liquidated damages to show that the schedule was not feasible, i.e. checking the schedule does not seem to be everywhere part of the contractor's legal liabilities.

(*h*) *Completion*

When the work is 'substantially complete' the engineer certifies that this is so; usually half the retention money is paid to the contractor at this stage; and the maintenance period, whose length is defined by the contract, begins. (The significance of the expression 'substantially complete' is that the certification may not be withheld because of trifling omissions by the contractor, although these may be reflected in deductions from his payments.) At the end of the maintenance period, if the deterioration in the work is no more than would be expected from fair wear and tear, the maintenance certificate is issued by the engineer, final payments made, and the contract is then complete; if the deterioration is excessive, the contractor has to make good the defects before final settlement.

7.5.2. Contractor's Reaction to Contracting Procedures

Even in developed countries, civil engineering contracting is a hazardous business: a study (quoted by Clough[2]) by Dunn and Bradstreet in the USA showed that such contractors are more failure prone than are other types of business, and that incompetence and inexperience are the major root causes of failure.

This situation—certainly as far as lack of experience is concerned—must be worse in many Third World countries, where the middle-aged professional with twenty years experience on major contracts is still quite a rarity. This points, yet again, for the need for sound pre-qualification and tender evaluation procedures.

Once past the first of these hurdles, the contractor will receive the contract documents, including the Bill of Quantities; he will, it is to be hoped, read the instructions, but even then, he may decide not to comply with them, at least in the matter of how he calculates the rates he uses. There will always be legitimate reasons for differences in rates between contractors—different equipment, different replacement policies, different wage rates, and so on—but two sources of difference exist which are not legitimate. The first concerns the case where the contractor thinks he has spotted a serious underestimate of the quantity of one type of work: he may aim to cash in by over-pricing this. The second difference, commoner and more serious, has its roots in the cash flow pattern of contracting work (Fig. 7.2): typically, there is a net outflow at first, gradually building up into a substantial inflow as preparation gives way to paying work, and then dwindling away—possibly as far as a net outflow again—during the tidying up phase. Profits tend to follow a similar pattern. Now, the contractor is normally required to distribute his central overheads and profit evenly over all the works. However, he

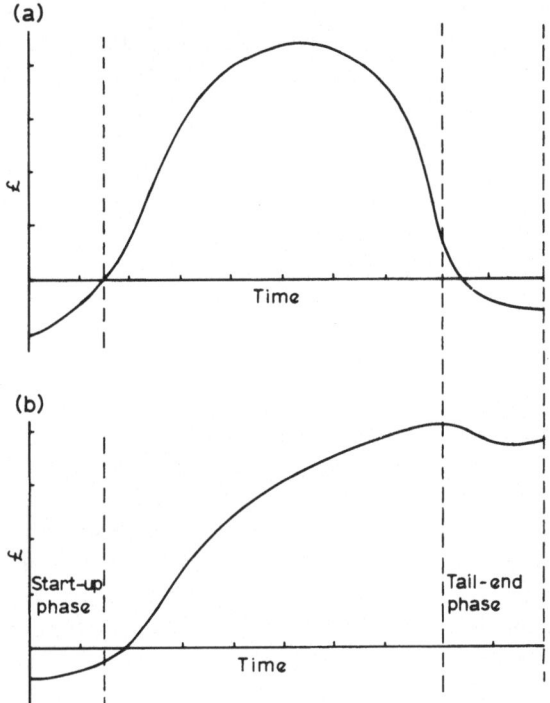

Fig. 7.2. Contractor's cash flow and profit. (a) Cash flow (monthly); (b) profit (cumulative).

will always be tempted to bring his cash inflows (and profit) forward: the early arrival of the money has value, as interest either earned or saved; and it may be required for financing other projects. He may therefore build into the rates for some of the earlier work part of his profit element, without necessarily increasing the total tender sum—this is called making an unbalanced bid.

Such bids are generally frowned on—and may even result in disqualification. With a competent and financially sound contractor, unbalancing may not cause too many problems: he will realise funds early, invest them soundly (probably as working capital on other projects), and be prepared for the fact that the tail-end period of the contract will almost certainly now need an input of funds. He has extracted a loan from the promoter, without the latter's consent and knowledge, of course.

However, a less sound and experienced contractor may well run into serious problems in the tail-end period: he may have underpriced the job, and simply not have the funds to bring in; he may have got the price right, but have used the funds which had been realised early for plant on another project which is not paying them back fast enough. (It is important to realise that it is not lack of profit that brings a firm down, but lack of cash to settle debts as they fall due—or rather, pass the latest date at which creditors can still be stalled.)

Problems now begin to snowball: the contractor is short of cash, so he delays payments to suppliers, who in turn, delay supplies; he fails to pay his workers regularly, and has trouble with everything from strikes to arson; and he switches effort to any activity which looks like generating cash, regardless of what this does to the overall project schedule. Soon, the promoter and contractor may find they are locked together in a ghastly embrace: the firm's principals may have a lot less to lose than the promoter if the contractor goes into liquidation, since it is usually very difficult to re-let the tail end of a contract, from which the most profitable work has already been stripped. It is possible to reach a situation where the promoter waives penalties, delays repayments on advances for plant, and even makes outright loans to the contractor, in a desperate attempt to get the project into some sort of working order. This is particularly likely to occur if the working pattern has allowed most of the funds to be drawn, with very little of the project being functional—this has happened on lined-canal irrigation schemes, where lining machines can do the straight runs, but junctions have to be hand-built, and are therefore left in the vulnerable tail-end period.

Add to this sort of situation a little unscrupulousness in the contractor's principals—especially if they are based outside the country— and a more sinister possibility arises: that, during the unprofitable tail-end of the project, they will yield to the temptation to abandon it, if the cost of all their liabilities exceeds the value of their bonds, plant, remaining work and retention funds. This usually also means abandoning their Agent to the mercies of an irate promoter!

7.6. IMPLICATIONS

The implications of the preceding sections for effective application of CPM-based project management systems are that:

(a) The planners must have the necessary resources to gather the data at a time when the obvious free sources may not be available.

(b) A provisional CPA of the project needs to be made before the preparation of contract documents for any work which is to be contracted out, and these should deal effectively with not only the currently visible critical interfaces between the areas of responsibility of the promoter, its sister organisations, and the various contractors, but with the fact that the project management system is a dynamic entity, and will require revision and adjustment. Section 7.6.1 deals with possible mechanisms for this, but it should be noted that these have to carry through to sub-contractors.

(c) The project sponsor's own project organisation structure should be designed so that it is responsive to the problems it will face.

(d) Intractable problems at the tail-end of each contractor's involvement should be minimised by all concerned parties pressing the promoter to accept that the engineer's role in screening out unsuitable contractors is crucial; and by inserting (where relevant) in the specifications some form of clear-up-as-you-go stipulation, so that, if there is a serious problem, as much of the project as possible is left in a usable condition.

(e) There should not be undue reliance on the force of law, the use of so called 'penalty clauses' to guarantee completion by a set date, as these are often least effective when most needed.

7.6.1. Possible Methods for Integrating CPM into the Contractual Structure

The central problem is that there has to be one, and only one, CPA for the project at any one time, and this has to reflect the true intentions and capabilities of all the agencies and parties involved, some of which may have a considerable degree of autonomy. Possible systems are:

(a) The owner engages an analyst, who attempts to engage the confidence of the contractors—and indeed of other agencies, such as other government departments, that are providing services. Their inducement to cooperate is that, if the conditions regarding the completion dates at crucial interfaces between the disciplines have been written into the contract fairly, the analyst is helping the contractor meet his liabilities, avoid delay, and avoid penalties. This is a common arrangement—however, the contractor may run into difficulties, and attempt to delude the analyst on, for example, the maximum feasible rate of progress, so that the analyst will need to liaise with the sponsor's engineers.

(b) A method advocated, but probably never used, is that the contractor is obliged to provide his own CPA services, and produce updated schedules. The difficulty with this is that there is no way of specifying what is an adequate level of provision; also, given that the analyst is the contractor's man, it is far more likely that he will be used—effectively—to delude the sponsors regarding whether or not crucial dates are going to be met, the reasons for delays, and so on. There is also the problem of reconciling the individual CPAs of the different agencies.

(c) As option (a), but with the contractor paying for updates. The hope here is that the contractor now sees some incentive to reduce updating expenses, by keeping the job on schedule. This is almost certain to be disappointed: the costs of CPA are small in relation to any cash flow problems the contractor may be struggling with, so that the result of using this approach would actually be to give the contractor a (small) incentive to conceal delays as long as possible.

Option (a) is still probably the best, provided that it is supported by the measures set out at the beginning of Section 7.6; it would also benefit from some form of familiarisation training in CPM for the staff involved.

7.7. PLANNING THE PLANNING

The actual process of conducting preliminary investigations and producing a plan is itself vulnerable to the same problems as project implementation—and CPM can make the same contribution to solving them.

A good example is provided by an investigative project with which the author was concerned; parts of three administrative districts were concerned, and the investigation was to determine the best siting of a predetermined number of deep irrigation tubewells (the number being related to the budget), what additional project components were needed to get the greatest possible benefit from the water, and to analyse the economics of the proposed investment.

A work plan had been drawn up—an intuitively produced bar-chart, the relevant parts of which are shown in Fig. 7.3. The staff involved were:

(a) Bar-chart (intuitive)

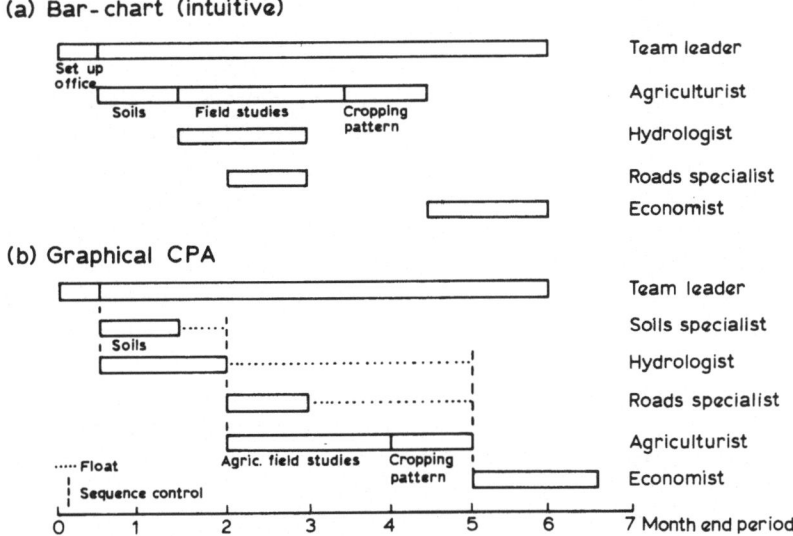

Fig. 7.3. Bar-chart and graphical CPA of tubewell investigation.

a team leader, experienced in this sort of work; a hydrologist, who was to review the existing data on groundwater reserves, to check that the total number of wells and total water abstraction were not excessive; a roads engineer with experience of transportation economics, who was to investigate the additional roads needed to deal with the extra traffic generated; an agriculturist, who was to review the existing soils data, determine the need for other project components and produce projected cropping patterns for an economist, who, with the team leader, was taking responsibility for report production.

The bar-chart showed work dates to the nearest half-monthly period (Fig. 7.3(a)); all went well during period 1—setting up the offices, getting transport, and so on. Periods 2 and 3—when the main activity was the soils data review, also went quite smoothly. At the beginning of period 4, the hydrologist arrived, and began his researches; and the agriculturalist began a survey of existing pilot schemes, to identify problems, as a guide to what additional project components might be needed. Logistical problems made it necessary to concentrate this survey in a few sub-districts, and for want of any better criteria, those with high concentrations of good soils were chosen. Shortly after, the roads man arrived, and organised his studies along similar lines, concentrating on sub-districts with good soils and poor roads.

It wasn't until the end of period 6, the end of the third month, that it became apparent that there was a problem, resulting purely from the phasing of the work: the hydrologist was discovering that the ground-water resources were concentrated in areas which did not have the highest concentrations of good soils. Unfortunately, they were also concentrated in areas with rather different agricultural systems and communications problems from those investigated by the other two specialists, so that their results were difficult to adapt to the areas in which the majority of the wells were to be sited. As a result, additional time was required.

Had the analysis in Fig. 7.3(b) been carried out, the total planned duration of the assignment would have been no longer than the actual duration turned out to be, and the actual staff time input would have been less.

REFERENCES

1. Institute of Civil Engineers (1971). *Civil Engineering Procedure*, Institute of Civil Engineers, London.
2. Clough, R. H. (1969). *Construction Contracting*, Chapter 1, Wiley, New York.
3. Clough, R. H. (1969). *Construction Contracting*, Appendix, Wiley, New York.
4. Handy, C. B. (1981). *Understanding Organisations*, Penguin Books, London.

CHAPTER 8

Using the Network

8.1. INTRODUCTION

We have seen that, in addition to a rational method of setting targets, any project management system needs a method of delivering instructions to the people responsible for the execution of project activities, and feedback—the sensory system through which management gains information about what is going on.

Both these components relate to the passing of information, upwards or downwards. If they are to function well, it is essential that they incorporate some sort of filtering device, so that the quantity and type of information reaching each person in the system is appropriate: too little, and he will be unable to act; too much, and he will possibly not be able to see the problem. For example, a field officer needs to know fairly precisely what he has to do, and when; a copy of the general specifications of the project both contains too much information, and is not sufficiently specific. At the other end of the scale, when progress reports reach, say, Deputy Secretary level, they should have been digested to the point where problems requiring action at that level (or above) stand out clearly from a brief summary of where the project is now, and how much its duration and resource requirements are likely to drift from the original targets. Inappropriate quantities and type of information are typically most serious at the higher levels in the project's sponsor's organisation, where points that require action are buried in a mass of trivial details that could and should have been resolved lower down in the hierarchy (using the word in a wide sense, to include other types of organisation structures than the 'Greek temple' format). The ideal flow of information from the project manager can best be represented by the sort of shape shown in Fig. 8.1—wide at the centre, i.e. at the project

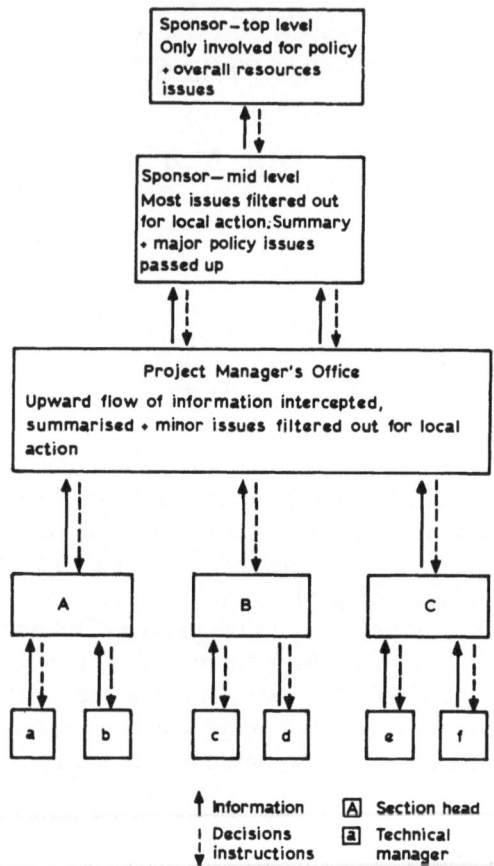

Fig. 8.1.   Information flow in a large project.

management team's level, and tapering upwards to a single point in the sponsoring agency, and downwards into several points, at the field level.

This chapter will deal, in turn, with: the downward flow of instructions; reporting back to project management; processing these reports; production of summaries for transmission upwards; and the special problems of financial reporting and control of expenditure.

For simplicity, we will be building the example for this part of the discussion around our first example, the watercourse improvement project (Table 1.1); the complete CPA of this project is set out in Table 8.1, and the network is reproduced as Fig. 8.2 for ease of reference.

Table 8.1. **Critical Path Analysis of Watercourse Project**

No.	Description	Follows	Precedes	Duration	Early Start	Early Finish	Late Start	Late Finish	Free float
1	Approval	—	All	12	1	12	1	12*	—
2	Premises	1	7	6	13	18	15	20	2
3	Trainees	1	7	6	13	18	15	20	2
4	Equipment	1	7	4	13	16	17	20	4
5	Trainers	1	7	8	13	20	13	20*	—
6	Watercourse selection	1	10	10	13	22	19	28	6
7	Office training	1–5	8	4	21	24	21	24*	—
8	Field training	1–7	10	4	25	28	25	28*	—
9	Construction materials	1	10	4	13	16	25	28	12
10	Start construction	All	—	1	29	29	29	29*	—

*Critical activities.

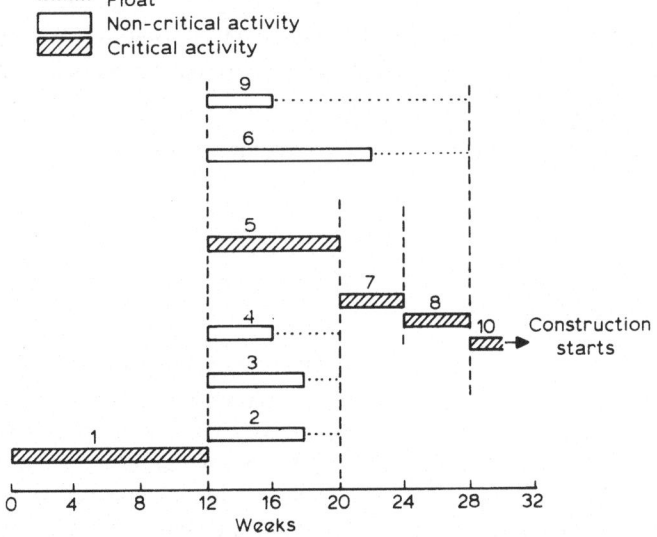

Fig. 8.2. Bar-chart of watercourse project (see Table 8.1).

Critical activities are indicated by an asterisk in the table; and, to keep the arithmetic tidy, the last activity, starting construction, has been given a false duration of one week. For the purposes of this chapter, it is assumed that there are no limiting resource constraints.

Only one other general point needs to be made now, but it is an important practical one. There will almost always be some generalised project reporting system in existence; the reporting phase of the project management system must be made compatible with the existing report formats (such as those shown in Fig. 1.1), with a minimum of transcription and rearrangement of the data. This not only saves effort, but helps gain acceptance for the improved method, simply by acknowledging the authority of the present arrangements. Also, if there is a traditional reporting system, running in parallel with a CPA-based system, it is likely that the two will diverge at some point, and there is then the danger that interested parties will pick and choose which set of results to believe on any particular occasion!

8.2. THE DOWNWARD FLOW OF INFORMATION: INSTRUCTIONS AND ADVICE

8.2.1. The Channels for the Information

At the beginning of the chapter on collecting the data, we drew up a sort of skeleton project, to indicate the sources of information. We can now put some flesh on the bones of that skeleton, to make it more life-like. The typical large project will consist of:

— a sponsoring agency, such as a government department of agriculture, and its own employees;
— sister organisations, such as a PWD, irrigation department, and so on;
— contractors;
— their sub-contractors.

Within these, there will be two main levels: section heads, responsible for major groups of activities, and technical managers, responsible for fewer, closely related activities. It is important that the terminology used doesn't blind us to the fact that many of the latter may have little familiarity with forms, and even bar-charts; in our context, many very effective men in this group may be illiterate, and this has to be taken into account.

If the advice of Chapter 7 has been taken, there will be adequate channels of communication down through the sponsoring body, and through the contractors and sub-contractors, and through the sister organisations. However, any instructions crossing the two barriers—sponsor/sister and sponsor/contractor—must have the agreement of the relevant senior people on the receiving end (e.g. the contractor's agent) on both its form, and the method by which it will be distributed.

Within the sponsor's own employees, and the sister organisations, the situation is simpler: there will usually be existing channels for the distribution of paper, orders, etc. However, it will still be worthwhile to enlist the sympathy and support of the various people controlling these channels, so that either they pass the material on quickly, or they accept its direct distribution by the project manager. In the latter case, the senior man would receive copies of the material: often his juniors are effectively on full- or part-time secondment to the project, while he himself is only involved to a small degree, and he needs to know the project's demands on their time. This situation could easily arise in our watercourse example, with some sub-district agricultural engineers being assigned full-time to the project, but being controlled by a district agricultural engineer, who is also heavily involved in general administration, running workshops, procurement, and so on.

Projects come in all shapes and sizes, ranging from the large and complex down to undertakings in which the project manager is also section head and technical manager, and there are no outside organisations involved at all. At the latter extreme, the channels of communication become unimportant—but the various checks and records described below remain useful. The rest of this chapter, in general, allows for the more complex situation: users of this book, with a simple problem, will readily adapt the principles to their case.

8.2.2. Who Needs What: Section Heads

In virtually all cases, it will be necessary for the section heads to receive fairly full details of the work under their charge. They have to make day-to-day adjustments in the distribution of effort between activities, to cope with minor problems, and they must be able to gauge the effect of the adjustments they make on the overall progress of the project. It will always be useful to provide section heads with a summary bar-chart of the whole project; this would be updated relatively infrequently. In addition, there would be their working instructions: depending on the quality of the staff available, these could consist of the relevant parts of

the network (in bar-chart form) and schedules, including the available
floats; alternatively, a simple bar-chart, with a priority grading of activities might be provided. Such a grading might group activities into:

— first priority: critical jobs, to be started and completed on the dates
   shown if at all possible;
— second priority: jobs to be started on the dates shown, if possible,
   but having sufficient float to suffer a slippage of, say, one week
   without unduly disrupting the project;
— third priority: jobs on which a slippage of, say, three weeks could
   be tolerated.

In both the latter cases, of course, the slippage is only tolerable if it
results from diversion of efforts to higher priority work, or from uncontrollable outside influences: as we have already seen, using up float early
in the project reduces the margin available later to cope with unforeseen
problems.

The instructions would normally be revised on a rolling basis, e.g.
monthly for the following three months.

8.2.3. Who Needs What: Technical Managers

Often, these men—who would often recognise themselves as foremen, or
by some other such title, rather than as 'managers'—will be fed instructions by someone outside the project manager's direct control. For
example, on a small milk processing plant project, the installer's technical
manager would be instructed through his agent, who would receive his
advisory 'instructions' as though he were a section head.

When the project manager can choose the format of these instructions,
there is a lot to be said for providing them in bar-chart form: these are
easily read because the bars can be labelled with symbols either instead
of, or as well as, lettering; and the bar-chart can be used as a diary, on
which the number of days worked is marked. The decision on whether to
produce individual mini-charts for individuals, or to issue copies of an
all-project chart, with the individual's responsibilities clearly distinguished in some way, depends on the size of the project, and the
drawing facilities available. In either case, the rolling revision method,
with a new chart for the next (say) three months, being issued every
month, should be used.

A technical manager needs to know:

— what activities he is responsible for;

Date Issued _____ Period Covered _____
Next Reporting Date _____

Activity No.	Description	Start on	Finish on	Actual days worked	Percentage physical completion	Forecast completion date		Remarks	
						A	B	Outward	Inward

Fig. 8.3. Instruction/progress report form.

— when they are to be started;
— when they should be completed;
— what other activities (outside his control) either they depend on, or depend on them.

In most cases, it is preferable for the technical manager not to be aware of the float available, as it is almost impossible to educate all project staff on the dangers of misinterpreting float.

8.2.4. Format of the Instructions
One good format—tried and proved in practice—for conveying instructions down to section head level, is the dual-purpose instruction and report form (shown in Fig. 8.3).

The forms would normally be prepared by the project manager (or his staff, on a large project) and issued to the section heads on the rolling revision basis already described; the precise frequency of revision depends on typical activity durations, but monthly is common, and usually fits in with administrative requirements for reporting. This arrangement gives the section head a view of where he is going in the medium term. The forms would also normally be accompanied by a bar-chart; either this, or the technical manager's bar-charts, functions as the diary on which actual work is recorded.

Before issue, for each activity, its number (agreeing with that on the network), brief description (explicit enough to ensure the recipient knows what is to be done), and early start and finish dates (described only as 'start' and 'finish') are inserted. Any additional information (e.g. priority grading) is inserted in the 'remarks outward' column.

Two copies of the report are issued, one for return to the project manager at the end of the period, one for retention by the section head.

8.3. THE UPWARD FLOW OF INFORMATION

8.3.1. Completing the Report Form
The actual number of days worked on each activity is taken from the bar-chart, and the percentage physical completion estimated. (This is only possible if activities are physically homogeneous, which is why that characteristic was included in the definition of an activity in Section 3.3. There may be problems in making the estimate for administrative activities, for which strictly only 0% and 100% completion are possible—

but even a comment in the remarks column, on the 'feel' of how the activity is progressing may be useful.) Percentage completion estimates would usually be the responsibility of the section head, working in collaboration with the technical manager.

If it is at all possible, the forms should be checked in the field by a member of the project manager's team, and brought in by him. This active collection has several important advantages over passively waiting for delivery:

— it ensures that all the report forms arrive back on time;
— it increases the staffs' consciousness of the fact that they are part of a project, that their efforts matter (and that their omissions may be noted);
— it enables accidental errors in completing the reports to be weeded out at an early stage;
— it enables misleading information to be detected, by allowing time for a visual check on progress.

The last item is often needed, more often that one would expect: commonly, project staff will put off the moment at which it must become apparent to the sponsor that some unacceptable delay is occurring. This is usually done in the hope of being able to retrieve the situation before that happens—a hope which is seldom realised.

Next, the forecast completion dates are calculated. If 25% of an activity is completed in 10 working days, then the remaining 75% should take

$$\left(\frac{100-25}{25}\right) \times 10 = 30 \text{ days}$$

This duration is converted to a date, by making the obvious allowances for non-working days, and entered on the form (Fig. 8.3) as forecast completion date A. This date may be misleading: if there are factors known to the section head that could either accelerate progress, or retard it, these would be incorporated into forecast completion date B. The reasons for entering this alternative date are put in the 'remarks inward' column; clearly, it is better if all this is done with the agreement of the project manager's representative, on the site, where physical problems can be inspected.

The form is now sent back to the project manager's office, with one copy being retained, for reference, at the site. Notes on any revisions required to the logic of the network are also collected and returned.

8.3.2. Processing the Form

The object of the processing is two-fold: to make any revisions needed at project level to the proposed plan of work; and to detect any issues on which it is necessary to seek authority or policy decisions from the higher levels of the sponsor's organisation. The latter is the subject of the next section.

Any revisions of the basic structure of the network will always have to be dealt with on an individual basis; the amount of work needed to incorporate them, and the seriousness of their consequences, can vary enormously. In many cases, they result from a failure of the agencies involved to think out carefully how they were going to physically perform the work, before becoming involved, and such a failure can result in very serious delays. The more novel the work and working methods, the more likely this is to occur.

In the commoner situation, where the activities and their inter-relationships have not changed, but the durations have, there are four possible effects of delays:

— the delay has affected a critical activity, and will produce a similar delay in project completion;
— the delay has increased the duration of a non-critical segment, by an amount greater than its free float, so that it now becomes critical;
— the delay has increased the duration of a non-critical segment by more than its free float, but it has not become critical because of delays on the original critical path itself;
— the delay has increased the duration of a segment by less than its free float, with negligible effects on overall duration.

The processing method has to distinguish among these, and produce: a revised programme of work, together with any necessary directions on the use of extra resources or different working methods; a revised forecast of the dates of major events of interest in the remainder of the project (including, of course, completion); and a corrected version of the network. From this information, it will also be possible to revise the expenditure proposals for the rest of the project. If the disruption is sufficiently widespread and serious, it may even be necessary to re-work the resource-scheduling calculations; the resource-use profile generated at the outset will indicate whether this is likely to be necessary, by showing which resources are most nearly used up at different stages of implementation.

A convenient method of processing the report forms, and distinguishing among the different types of delay, is to transfer them to a segment delay check form, such as that shown in Fig. 8.4. The activities are grouped into segments, since free float is a property of the segment, not individual activities; individual critical activities can be treated as segments (since they have no float anyway). A tick is inserted in the 'free float exceeded?' column if the delay exceeds the free float; this also means, of course, that a tick appears in that column if there is any delay on a critical activity/segment. The appearance of any ticks in that column indicates a need for the re-working of the forward and backward pass calculations; sometimes, inspection of the network will enable the corrections to be made intuitively, but a complete revision is safer, because, otherwise, a stack of intuitive adjustments may accumulate in successive reporting periods, and conceal serious errors.

Detecting and allowing for early completions of activities is not so important: they will never have any effect (except through releasing scarce resources for use elsewhere in the project) on overall duration unless they affect critical activities. Even in that case, since there is often little float left in the network at the end of resource scheduling and crashing, an accelerated finish on a critical activity may not shorten project duration by very much—since there will often be a parallel near-critical activity which will take over as the controlling factor. Even in our watercourse project example (see Table 8.1), which has more float (and fewer parallel near-critical segments) than most real-life networks, reducing the length of the critical activity 'trainers' (activity 5) by three weeks would only shorten the project by two weeks, because the activities 'premises' (activity 2) and 'trainees' (activity 3) would both then become critical as soon as the length of activity 5 fell below six weeks.

To demonstrate the method of revision, we will look at our watercourse project, at the end of week 19, when the following had happened:

— activity 1 (approval) had been completed on time;
— activity 2 (premises) had been completed two weeks ahead of schedule;
— activity 3 (trainees) was over-running seriously;
— activity 4 (equipment) had been completed on schedule;
— activity 5 (trainers) was apparently on schedule;
— activity 6 (watercourses) was over-running seriously;
— activity 9 (materials) was completed;
(activities 7 and 8 were not yet due to start).

Activities		Segment					Free float exceeded?
Numbers	Description	Late finish	Forecast finish	Total delay	Free float		

Fig. 8.4. Segment delay check form.

None of the critical activities had been delayed. Of the delays to (currently) non-critical activities, the delay affecting the trainees originated in administrative problems, so that the forecast completion date calculation could not be made, and only the forecast estimate B on the report form (Fig. 8.3) could be made. The other delay—to watercourse selection—had resulted from difficulties of travel and of consultation with local leaders, that were expected to persist; only 30% of the target number of watercourses had been identified in 18 working days, so that the estimated time to complete the remaining part of this activity was:

$$\frac{100-30}{30} \times 18 = 42 \text{ working days} = 6 \text{ weeks}$$

(Calendar days were considered to be the appropriate unit for this item.) This brings the finish date to the end of week 25 (19 + 6); Table 8.2 shows the appearance of the completed report form.

The information from Table 8.2 is transferred to the delay check form, which is shown as Table 8.3 (note that here, our segments each consist of only one activity).

Note the difference between the two delays: despite the fact that the delay in the watercourse selection activity is bigger, it has virtually no effect, being within the available free float.

The next step is to update the network, which the appearance of a tick in the 'free float exceeded?' column proves is necessary. To produce a completely revised network, the following steps are needed:

— insert the actual durations of the completed activities;
— insert the expected total durations of ongoing estimates;
— insert the current forecasts of the durations of activities yet to start;
— re-work the forward and backward pass calculations.

Alternatively, a revision of the network for only the remaining part of the project may be made, by carrying out the calculations using the anticipated remaining durations of ongoing activities, and the anticipated durations of those yet to start. In Table 8.4, the former option has been chosen; the table shows that project duration has been extended by only one week. It would not have been possible to determine this true effect of the delays by any other means than CPM, and there could have been needless panic over the trainee secondment issue.

The fact that, in this example, the critical path has moved should not be a surprise: there are a number of parallel paths in the network, in the

Table 8.2. Progress Report for Watercourse Project
Issued: Week 16   Reporting period: weeks 15–19

Activity		Start	Finish	Days worked	Percentage physical completion	Forecast completion date		Remarks	
Number	Description					A	B	Out	In
3	Trainees	(13)	18	NA	—	—	21	—	Difficulties experienced in getting trainees released by seconding depts
4	Equipment	(13)	16	NA	100%	—	—	—	—
5	Trainers	(13)	20	NA	—	—	20	—	No problem expected
6	Watercourses	(13)	22	18	30%	25	—	—	Travel problems and difficulty of getting local leaders to agree to participate
9	Materials	(13)	16	NA	100%	—	—	—	—

All dates shown are weeks from the start. Dates in parentheses refer to ongoing work. NA = not applicable. Activities 1 and 2 are already completed; activities 7 and 8 are not due to start in this period.

Table 8.3. Segment Delay Check Form for Week 19 of Watercourse Project

Activity		Segment				Free float exceeded?
Number	Description	Early finish	Forecast finish	Total delay	Free float	
3	Trainees	18	21	3	2	√
4	Equipment	16	16	—	4	—
5	Trainers	20	20	—	—	—
6	Watercourses	22	26	4	6	—
9	Materials	16	16	—	12	—

relevant area (activities 1–5 in Fig. 8.2) of similar length. Recalling the material presented earlier on statistical considerations, we can see that it was always quite likely that this sort of movement would have occurred. The proper approach to using the results of a CPA under these circumstances is to anticipate which activities may become critical during implementation, weigh up the costs of providing some form of stand-by resources to cope with this eventuality and be prepared for the movement. Treating the initial CPA as a fixed framework is almost always a mistake: it is a thing which must evolve during implementation. Failure to realise this explains some of the discontent seen in developed countries—particularly in the construction industry—with CPM, where any drift of the actual schedule and critical path from those originally planned is seen as a failure of the technique. The real world just is not so controllable, that plans can be made and executed with no adjustment. The object of a good project management system is to identify or, if possible neutralise, deviations from plan before they can spread a ripple— or even a tidal wave—of disruption through the project.

The above analytical procedure may seem a little elaborate when applied to our simple example—but the author has seen a number of cases in which some activity, apparently a side issue to the main technical activities (such as staff housing, getting processing plant operators trained) eventually became a major critical delaying factor, simply by persistent neglect of the seemingly unimportant. A good CPM-based system would have spotted those activities, quietly working their way to the front of the field, like an unfancied horse in a race—and stopped them! The procedure described counteracts the tendency of many project managers to perceive as important only the main technical activities.

Table 8.4 Revised CPM for Watercourse Project at Week 19

Number	Description	Follows	Precedes	Duration†	Early Start	Early Finish	Late Start	Late Finish	Free float
1	Approval	—	All	12	1	12	1	12*	—
2	Premises	1	7	4	13	16	17	21	—
3	Trainees	1	7	9	13	21	13	21*	0
4	Equipment	1	7	4	13	16	18	21	—
5	Trainers	1	7	8	13	20	14	21	1
6	Watercourses	—5	10	14	13	26	16	29	3
7	Office training	7	8	4	22	25	22	25*	0
8	Field training	7	10	4	26	29	26	29*	0
9	Construction materials	1	10	4	13	16	26	29	—
10	Start construction	All	—	1	30	30	30	30*	0

*Critical activities.
†Revised values.
—Free float not shown for completed activities.

Incidentally, it is possible to get some idea of how likely it was, from the outset, that the shift of the critical path in our example would occur. Consider the group parallel activities (2–5) containing the original critical path; longest of these was 'trainers' (activity 5), planned to take eight weeks, and the second longest, either 'premises' (six weeks), or 'trainees' (six weeks). The rule of thumb suggested in Section 6.3 was that change of critical path might occur if the difference between the lengths of the paths to the ends of such a pair was less than the standard deviation (SD) of the more variable of them. In other words, you could expect movement of the critical path if the SD of the total duration of either activities 1 and 5 or 1 and 3 was more than two weeks. Does this seem implausible? Well, in the former case, if the duration was sufficiently variable to make 50% of occurrences fall outside the range 18·7 to 21·3 weeks, this should have made us expect such a shift (remember, for a normally distributed variable, 50% of occurrences lie within 0·67 SD of the mean). That amount of variation certainly doesn't seem impossibly large.

8.3.3. Progress Reports

At project manager level, these would be fairly full; for example, a report at week 19 on our watercourse project would cover the points that:

— equipment and materials procurement had occurred on schedule, and secondment of trainers was proceeding smoothly;
— that there was some delay in watercourse selection, and that, although steps were being taken within the project to prevent it increasing further, by better pre-identification of watercourses by discussions with local government officials, the delay was not likely to have serious consequences;
— secondment of trainees was delayed by circumstances outside the project manager's control;
— progress is generally reasonable, but that a small delay in completion date is now expected, because of the last point.

In reducing this to a summary, the best discipline is to ask the questions 'Who am I reporting to?', 'What does he need to know?', and 'What do I want back from him?'

In the case of a project like ours, the reports would probably be going to the regional or provincial agricultural officer. He doesn't really need to know detailed adjustments inside the project, and the thought that 'it can do no harm to tell him' should be strangled at birth: if everyone takes the easy route, by not sifting their reports carefully, the unfortunate man will suffocate under the weight of paper. Also the project manager will have decided that, in this case, what he wants back is help and support to get his trainees; he won't get them unless his request stands out clearly from a plain and simple background. The project manager's monthly report on this period to the regional official would therefore indicate that:

— progress on procurement of equipment and materials, secondment of trainers and watercourse identification is generally good;
— there has been a problem with secondment of trainees and this will delay project completion slightly;
— action is requested, to bring pressure to bear on the heads of the seconding departments to release the trainees promptly to prevent this delay growing.

The regional official himself would be producing an even more concentrated summary, possibly for the Ministry of Agriculture, indicating only that progress is generally good, and steps have been taken to contain a problem which had arisen, so that only a minimal delay is now expected. In more complex cases, and with greater delays,·a schedule of revised dates of *major* events would also be relevant. At that level, the report will be addressed to officials with limited contact with agriculture in the field, who not only have a greater chance of drowning in information (because more people report to them), but have a lot less interest in, and possibly understanding of, the technical details anyway.

Probably the best single pruning technique is to add an additional question to the three already suggested: 'Which of the current problems should I be resolving myself, instead of passing them up to the next man?' Certainly the worst possible reporting practice, from the information management point of view, is to produce a single compendious report on the project, which is distributed to all who require any report at all: it wastes the time of those readers who have to winnow out the important from the trivial, and reduces the project's ability to drive home a clear message regarding what help it needs—or even, what praise it deserves—to the more senior recipients.

It is in the process of reporting, and requesting essential action, that the issue of an appropriate organisational structure arises: the better, more appropriate it is, the faster the turn-round of requests for action into policy decisions and help. Anything that can be done to make the structure more appropriate at the planning stage will accelerate this turn-round. However, if a matter is serious enough to really need referring upwards, a very rapid response is unlikely; a reasonable expectation would be to get a reply in time to incorporate its effects into the following set of instructions.

8.3.4. Visual Progress Records

Project managers and responsible officials in the sponsor's organisation derive great psychological satisfaction from visual progress charts; they don't really add much extra information, but may give a better 'feel' for the state of the project.

A very good format is a bar-chart, with the activities grouped into bands according to the agency involved; critical activities are distinguished, e.g. by a heavier outline around the bars; progress is indicated by shading a proportion of the bar for the appropriate activity; and today's date is indicated by a vertical cursor. Figure 8.5 shows this type

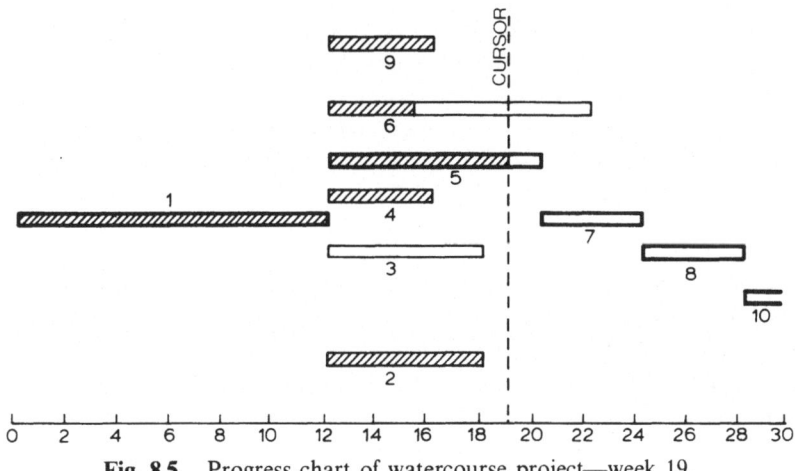

Fig. 8.5. Progress chart of watercourse project—week 19.

of chart, drawn for the case we have just analysed. In this form of presentation, unshaded bars to the left of the cursor represent delays; and unshaded areas to the left of the cursor on the critical activity bars represent serious delays. The interpretation of shaded areas to the right of the cursor is a little more difficult, since it appears, at first sight, that these should never occur. Apart from simple mistakes, there is one quite innocent explanation: the schedule depends on resource constraints, and in Chapter 4, we saw that once critical activities have been supplied with a resource, the order of allocation to other activities is somewhat arbitrary; as a consequence, so is the order in which these other activities start. In this case, relatively minor considerations of ease and convenience may legitimately alter the order of starting the activities.

Rather more sinister is the possibility that the network no longer reflects the true intentions of all the parties on exactly how the job is to be done. This can be the result of a simple realisation that the proposed method, as it was inserted in the network, is not feasible, or not the best method; particularly if the agency responsible for the activity is a contractor, and the alteration reduces the chances of his meeting any key dates for completion of parts of his work which have been written into the contract, the Agent should be alerted to the fact. In any case, it is essential to make sure that there is only one network on the project, the same one in the field as on the wall of the office.

The worst case in which progress-shading appears to the right of the cursor is where the network is realistic, and work is being done out of sequence, with the result that remedial work will be required. This would happen if, for example, road surfacing was being laid in an area in which heavy bush clearing equipment was still being operated. This sort of situation can result simply from carelessness, but it always needs investigation: it could be a symptom of a contractor running into cash-flow troubles, and bodging work to generate payments now, at the risk of increased problems later.

There is an alternative method of marking progress called 'progress laddering' which is easier to use if, instead of drawing the chart, one of the proprietary systems using perforated boards and plug-in plastic segments is used. As work is completed, the appropriate proportion of the bar is detached, and set vertically, like a ladder, in a prepared place, at the side of the chart. This really only works with highly aggregated activities, is visually more cluttered, and provides a less clear warning of work which is being done before it is due, than the bar-chart and cursor method. In general, the proprietary systems are not as good as drawings: parts get lost, or deteriorate in extreme climates, and the system is vulnerable to idle fingers in a way that a drawing is not (you can readily see graffiti!).

8.4. FINANCIAL REPORTING

8.4.1. Expenditure Curves—Principles

The heavy emphasis on financial information as primary data in traditional project management systems almost conceals the fact that it is physical activity that determines expenditure. (The only other influence is, of course, price changes.) The CPM approach rejects financial data as the primary item, and builds up the financial targets and progress results from a sound analysis of the underlying physical activities.

The principle of building up the financial targets is simple enough: if you know the timing and the cost, you know the time-pattern of disbursement. However, there are variations in the timing of the payment relative to the start of the activity. Some of the most important types of pattern of payment are:

(a) Simple lump sum payments, at some fixed point, relative to the start of the activity. Purchase of equipment will usually result in a payment either at the end of the activity, or after the lapse of the necessary period for billing and processing of the payment by the sponsor; in some countries, work carried out by PWD-type organisations for a sponsor is paid for, in its entirety, at the outset (or even in advance).

(b) Staged payments. This may well occur in the case of imported plant, where a gradation of the payment such as the following may occur: 20% on signature of contract; 50% on clearance through the local port; 25% on commissioning; 5% at the end of six months' satisfactory operation.

(c) Payments for civil engineering works (Bill of Quantities Contracts). As described in Chapter 7, these are paid for as work is done, except that a proportion of the payment is withheld in the retention fund, and released in two instalments, one at the point of substantial completion, and one at the end of the maintenance period; and there is the inevitable delay between certification of the amount of work done and release of the payment.

(d) Payments for lump sum and turnkey contracts. The payment terms for these—particularly the latter—will vary greatly, and no general rule can be applied safely.

(e) Recurrent costs.* Some activities will generate a steady stream of costs—hiring labour is an obvious example. (But even here, a little care is needed: the cash outflow will consist not only of wages and salaries, but must also include any welfare payments, transport and housing costs, etc., paid by the sponsor.)

*'Recurrent' is used in its elementary dictionary sense; in some countries the phrase 'recurrent expenditure' defines a category of funding, and may include capital items, such as purchase of equipment.

Table 8.5. Calculation of Cost Curves for Watercourse Project: Example Work Sheet

Number	Description	Payment (by weeks),										
		1	2	3	4	5	6	7	8	9	10	11
	Target—all activities at early start											
1	Approval											
2	Premises											
3	Trainees											
4	Equipment											
5	Trainers											
6	Watercourse selection											
7	Office training											
8	Field training											
9	Construction materials											
	Total											
	Monthly total				nil				nil			
	Cumulative total											
	Target—all activities at late start											
1	Approval											
2	Premises											
3	Trainees											
4	Equipment											
5	Trainers											
6	Watercourse selection											
7	Office training											
8	Field training											
9	Construction materials											
	Total											
	Monthly total											
	Cumulative total											
	Actual expenditure to end of month 6 (week 24)											
1	Approval											
2	Premises											
3	Trainees											
4	Equipment											
5	Trainers											
6	Watercourse selection											
7	Office training											
8	Field training											
9	Materials											
	Total											
	Monthly total											
	Cumulative total											

*370 paid out in week 33/month 9.

12	13	14	15	16	17	18	19	20	21	22	23	24	25	26	27	28
						50				50				50		
						180				180				180		
				115												
								150				150				
	5	5	5	5	5	5	5	5	5	5						
									15	15	15	15				
													20	20	20	20
								370								
nil	5	5	5	120	5	235	5	155	390	250	15	165	20	250	20	20
				135				400				820				310
				135				535				1355				1655
						50				50						50
						180				180						180
						115										
						150				150						
							5	5	5	5	5	5	5	5	5	5
									15	15	15	15				
													20	20	20	20
							5	500	20	20	20	400	25	25	25	255
								505				460				330
								505				965				1295
								60								
								180								
				115												
											150					
				10	10	10	10	10	10	10	10	10				
								370								
				125	10	10	10	10	250	380	160	10				
				125				40				800				
				125				165				965				

*

(f) Mixed payments. Some activities will generate a mixture of lump sum payments and recurrent costs; this may happen with purchases of mechanical plant, e.g. tractors, or acquisition of premises for rent, where both rent and initial premium payments are made.

In drawing up financial targets (which are usually represented as graphs of expenditure against time), from the schedule of activities, there is a further point to be considered: the difference between early and late starts represents a delay which is not serious; the corresponding effect on the rate of expenditure should therefore not produce the conclusion that the project is drifting off schedule. There are, in fact, two target cost curves: both embody the considerations set out in (a) to (f) above but one cost curve assumes that all activities start at their early start dates, while the other cost curve (the lower of the two) assumes that all activities start at their late start dates. They enclose a zone of legitimate expenditure patterns on the graph of expenditure against time.

Any pattern of expenditure—represented by an actual cost curve—which lies between these two target cost curves is consistent with sound progress (in the sense that no irrecoverable delay has occurred). Unfortunately, if you know that the actual expenditure curve lies between the two cost curves, all that you are guaranteed is that your expenditure pattern is consistent with sound progress: it is consistent with a lot of other things, too! There is no way around the basic ambiguity that results from the fact that such an expenditure pattern might have resulted from under-expenditure on critical activities, and over-expenditure on non-critical ones. (A check for this is described later.)

8.4.2. Expenditure Curves Example
We can now build up target cost curves for the example re-introduced at the beginning of this chapter. Table 8.5 shows how this is done. The assumptions made in constructing this table are:

— no overheads (e.g. administrative salaries and expenses) are charged to the project;
— rent and salaries are paid at four-weekly intervals, starting from the end of the activity that results in their acquisition;
— watercourse selection and the two training activities result in cash expenses for materials, travel and accommodation that are paid immediately;
— payment for construction materials is made five weeks after the end of the purchasing activity, but other equipment is paid for at the end of the purchasing activity.

The monthly totals are shown in cumulative form in Fig. 8.6; the S-shaped graph of the early start expenditure curve is very typical, resulting from the sort of slow initial build up, main peak of activity and expense, and tailing-off period that we saw in the case of contractor expenditure (Fig. 7.2). The late start curve shows this less clearly, with the scale and intervals used disguising the initial build up.

Now, consider what the actual cost curve would look like at the end of week 24, if the following conditions apply.

(1) Approval had been delayed by three weeks.
(2) Acquisition of premises had been delayed by activity 1 and had cost 20% over budget.
(3) Secondment of trainees had been delayed by activity 1.
(4) Equipment purchase had been accelerated, so that the payment was made, at the price budgeted for, on the date originally scheduled.
(5) The appointment of trainers was delayed by activity 1.
(6) Watercourse selection was delayed by activity 1, and cost twice as much as budgeted for.

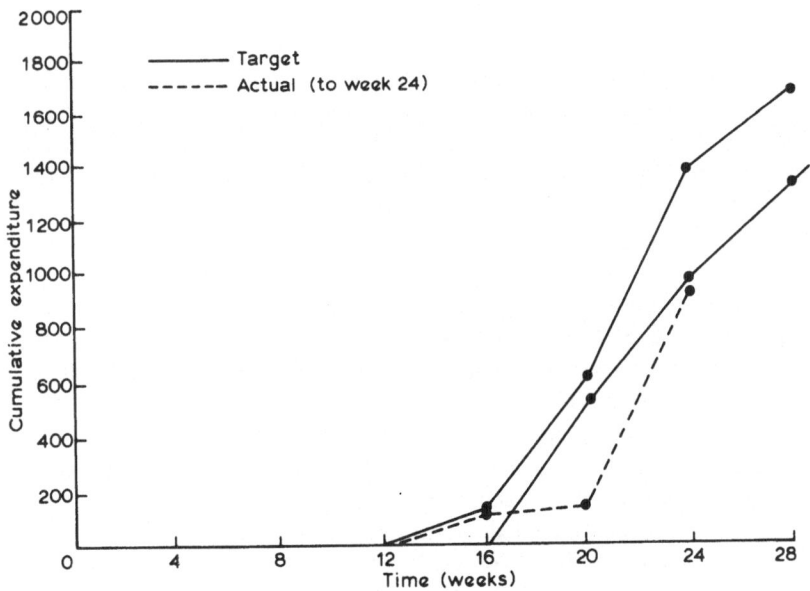

Fig. 8.6. Target and actual expenditure curves.

(7) Office training had not started, being delayed by activity 1.

(8) Field training had not been due to start yet, even on the original schedule.

(9) It had proved possible to bring the purchase of materials forward, into week 22.

In Fig. 8.6, the three cost curves have been plotted, with the target cost curves shown as solid lines, and the actual expenditure as a dashed line. The effect produced is totally misleading: although the initial lag shows up clearly, there is an entirely spurious impression of accelerating progress produced by the early acquisition of materials (which is quite a common reaction to underspending).

8.4.3. Financial Progress Reporting Forms

Figure 8.7 shows a specimen financial reporting form, from which material for standard traditional financial progress reports can readily be extracted. The summary section is particularly useful: a low ratio of actual expenditure to planned expenditure on critical activities would immediately suggest that something was amiss. For example, the summary section for the example of the last section, at the end of week 24, would correct the misleading impression given by the total expenditure figures and graphs, as shown in Table 8.6.

Just like physical progress reporting, financial reporting has to be subjected to a process of digestion and concentration. The results of this process, as the report passes up the chain of command, would be a reduction in the number of headings: the project manager's own report to, say, the Regional Agricultural Officer would probably detail all the items referred to; the latter's report to, say, Provincial Ministry level might well indicate either only one summary line for the whole project, or be broken down into major headings, consistent with the divisions in the financial sanctions (possibly staff, accommodation, travel, general expenses, and special items, such as the construction materials). Similarly, on an irrigation scheme, the project manager would prepare financial reports for his own use and for, say, a project steering committee, in fair detail, showing expenditure on earthworks, canal lining, installation of regulators and turnouts, field drainage, main drainage, and so on. However, he would have already aggregated the data himself—and the version intended for, say, the Deputy Minister of Agriculture might well show only irrigation works as a single item, with staff housing as another, agricultural development as a third, and so on; and, of

Date Prepared _____ Period: From _____ To _____

Indicate whether this is current report ☐ or forecast ☐ (Tick one)

Activity/ type of expenditure	Cumulative expenditure to start of period	Planned work			Actual work			Value of work to end of period	Tick if critical	Remarks
		Quantity	Rate	Amount	Quantity	Rate	Amount			

Cumulative Summary:

	Planned	Actual
Critical		
Non-Critical		
Total		

Fig. 8.7. Financial progress reporting form.

Table 8.6. Financial Summary for Example:
Critical versus Non-Critical Expenditure

	Planned	*Actual*
Critical	360	150
Non-Critical	605	815
Total	965	965

course, such reports could be made less frequently—probably at three- or six-month intervals, instead of monthly.

Any written commentary on the tables should be similarly tailored to its recipient's needs.

8.4.4. Financial Forecast Report

At relatively long intervals—probably never more frequently than quarterly—a revision of the financial targets for the remainder of the project's life should be made; this is a straightforward application of the cost-curve calculations set out above, incorporating any anticipated price changes and, of course, the latest up-date of the schedule. The object of this report is to alert the sponsors to major changes in the likely pattern of requirements for funds.

CHAPTER 9

Critical Path Analysis by Computer

9.1. DO YOU REALLY NEED A COMPUTER?

Some of the techniques described early in the book—especially resource scheduling—are fairly tedious; all (despite the existence of cross-checks) are susceptible to clerical error; and some of the more advanced techniques described in Chapter 6 require an amount of computation which really is beyond manual processing. Therefore, it is not surprising that many practitioners of CPM have tried to apply modern electronic computers to the problem, often with great success. However, there are serious limitations on their usefulness in the context of development projects. All the components—including the high speed input and output devices—are finely engineered, often fragile, and readily damaged by dust and humidity. As a result, under project conditions, computers suffer from severe drawbacks:

(a) Service and spares availability may be poor, and complicated by problems with import formalities for replacement parts. Therefore, a computerised project management system can easily be brought to a halt, while the project itself goes trundling on.

(b) The equipment may be exposed to unreasonably harsh operating conditions, with excessive dust, heat, humidity, and voltage fluctuations.

(c) Other essential support services may be lacking—notably those relating to user training and software support (i.e. making sure the programs actually work on your machine).

In addition, even if these problems can be overcome, there are advantages in manual calculation: the person doing it gets a far better feel for the state of the project than he would if he merely passed the whole lot on to the computer clerks.

However, there will be a limit to the amount of time that can be devoted to hand calculation, and, in a large, complex project, subject to revisions of the network, and involving substantial resource scheduling problems, computerisation should be considered. The verdict should only be 'yes' if certain key criteria can be met: a suitable environment, reasonably clean and dust-free, with some air-conditioning, must be available, as must proper support (discussed in Sections 9.3 and 9.4). In addition, of course, there must really be a need for computerisation: it is difficult to lay down hard and fast rules on the point at which manual processing ceases to be a reasonable proposition, as it depends on the reporting frequency, the extent to which the work is familiar to the agencies involved (both of which affect the potential frequency of network revisions), and the number of activities.

Table 9.1 indicates the levels of these parameters at which manual processing becomes sufficiently tedious for the error rate to increase noticeably.

Table 9.1. Manual versus Computer Processing

	Working method					
	Familiar			Unfamiliar		
Reporting period (months)	1	2	3	1	2	3
<100	(M)	M	M	(M)	M	M
101–200	(M)	(M)	M	C	(M)	(M)
201–300	C	(M)	(M)	C	C	C
300+	C	C	C	C	C	C

C = Computer processing advisable; M = manual processing possible; (M) = manual processing possible, but difficult if there are extensive resource scheduling problems.

Computerisation has admittedly become more attractive with the development of microcomputers: progressive miniaturisation has brought about a situation in which computing power that would have filled a large room, and needed full air-conditioning, 10 years ago, can now be fitted into a desk-top machine, which is relatively robust, and will work with no more than good office standards of cleanliness and climate control.

A complete system, including a really good program, will cost something of the order of £6000 (at late 1983 prices), about 55% of which is accounted for by equipment. This excludes any freight, insurance, or import duties to the project country, any training courses needed by staff (including travel and accommodation), air conditioning, and provision of a sufficiently stable power supply. The temptation to reduce that sum by allowing someone from either the project staff, or the sponsor's staff, to write programs, must be strenuously resisted: programming is notoriously liable to extensive overruns, and programs prepared by non-professionals tend to be difficult for people other than the programmer to use (see Section 9.4).

9.2. COMPUTER TERMS

A certain amount of information about computer terminology is necessary, to understand the subsequent discussion.

A computer is a calculator that can store a series of calculating instructions, and carry them out in sequence, provided that it can handle instructions like: 'if there are no more activities to calculate early starts and finishes for, begin backward pass calculations, otherwise begin early start and finish calculations for next activity.' It is the ability to execute commands in different ways, conditionally on the outcome of some internal test, that is the distinctive feature of computers.

Such a sequence of instructions is a program—you do not 'program in' data, you 'input' it, or 'key it in'. Programs and data are fed into the working parts of the computer via a keyboard, which is basically very like that on a typewriter. Programs can go in in three forms: as strings of numbers and letters (called machine code) which essentially are the numbers of the various functions which the computer can execute; as an assembly language program, where the instructions at least remind the programmer of the function (e.g. SUB instead of 031 for subtract), and are converted by a special program called an assembler to machine code; or as a high-level language program, in which several machine code steps are compressed into one statement, which is expanded into machine code by a special program called a compiler. (Most non-professionals use these high-level languages, e.g. FORTRAN, BASIC, for programming, because they are fairly readable, but many professionally prepared programs are in assembler or machine code.)

Programs are the main component of what is called 'software'—some people also include documentation, training manuals, and advice under this heading; 'hardware' is the actual equipment. Someone said 'hardware is all the bits you can actually kick, software is the rest'! Systems software consists of programs like assemblers, compilers, and the operating system, which handles the program's input and output, allocates working space inside the computer, and warns of serious error conditions generated by a user's program. (Unfortunately, one common operating system is called CP/M.)

Often, the software for a particular application—e.g. project management—is described as a package, because it consists of several fairly independent pieces of program, linked together.

In addition to the keyboard, the computer communicates with the outside world through other input/output devices, the usual ones being:

(a) VDU (visual display unit), like a small TV screen, for displaying messages and input before it is worked on. Many programs display a form for, e.g. the description of an activity, with spaces for duration, number of preceding activities, resource needs and so on. These spaces are filled by using special keyboard buttons to move an indicator to the appropriate space, and typing: the characters then appear in the appropriate space. When the form is filled, all the data items are sent off into the 'works' of the computer by pressing the keyboard button marked 'enter' or 'return'.

(b) Disks. On microcomputers, these are often referred to as diskettes, or floppy disks, to distinguish them from the much larger rigid disks used on big computers. They use magnetic storage, just like ordinary magnetic tape.

(c) Printer. Although there are graph plotters, most software uses the printer to produce items such as bar-charts, as well as text and tables.

There are other input/output devices, e.g. magnetic tape, and readers and punches for both cards and paper tape, but these are uncommon on microcomputers.

The distinction between microcomputers and the rest has already been made: they are merely computers which are desk-top instruments, and may be comparable in size with mainframe (see below for a definition of this) machines of 10 years ago. Size is measured in terms of the number of storage locations for programs and calculations inside the computer, usually in units of K, 1 K being approximately 1000 such locations. Size is often referred to as memory, storage, or core size.

'Mainframe' is almost as loosely defined a term as 'microcomputer'; essentially, it is a computer physically large enough to require its own sizeable room. From the user's point of view, one very important difference between the mainframe computer and the microcomputer is that—except in very modern installations—the mainframe computer is often only directly accessible to its specialist operators. This means that interactive use, in which the user can type material directly into (and react to messages directly from) the computer is not possible with the sort of mainframe equipment often installed as the central computing facility in a Ministry. This has such a profound effect on the ease with which data, for both the original network and updates, can be cleaned up and used, that this author, at least, would be very reluctant to rely on such an installation for project management services. (The issue of interactive use and cleaning-up data is dealt with in Section 9.3.1).

9.3. REQUIREMENTS OF A GOOD CPA SOFTWARE PACKAGE

It is important to realise that you must choose the software first, not the machine—although the choice of software may be restricted to that which will run on machines for which adequate support is available (see Section 9.4). There are a number of things required from a good program, some of them general, applying to all programs, and others specific to CPA programs; these are dealt with in turn.

9.3.1. General

(a) The package must, of course, work! This, perhaps surprisingly, is the easy bit: it is much easier to write a program that does the calculations correctly than to progress to the point where it is 'user-friendly', i.e. to ensure that the format for data input is readily understood and used by someone who has little or no experience, that mistakes caused by the user are intercepted before they cause problems, that the output is well presented, and so on. In the recent past, large numbers of people, in a wide variety of fields, have been tempted into program writing by the potential power of the equipment to solve large problems quickly, without realising the enormous amount of work needed to convert a program that will solve the problem—if used by its originator—into one which is not readily stopped by the sort of minor errors that other users will inevitably make. Not only is there a lot of work involved in this step, but it is very difficult for the programmer to check that he has done it thoroughly: he has to anticipate all the silly mistakes, and combinations of silly mistakes, that users may make, and protect against them. It is this latter part of programming that non-professional programmers find most difficult, and their failures with it are so common that, in general, it is unwise to rely on specially written programs. Nowadays, there is a large variety of good software available, at prices that (in the cases where a computer is really needed) are quite sensible.

(b) The software should have local support. This means that the originator of the package should be able to provide, at least in the same country, someone who can give advice and training, deal quickly and effectively with any problems of incompatibility between the software and the machine and its system software, and correct any errors that may remain in the program. Errors are rare with professionally prepared software: the originator's entire business reputation depends on such errors being removed before the software gets into the market. However, they do occur, and it is essential that they do not bring the project management system to its knees. Sadly, many users call in the software support people to deal with the results of their own errors, which they blame on either the equipment or the software!

The only alternative to good local support—and that for software from a reliable source—is to get software that is so well known and well proven that the risks of problems are minimised. This means that the project manager must be able to establish for himself that the package has been in extensive use, on exactly the same machine, without any problems; and, even then, it must be realised that some element of risk remains.

An important aspect of software support is training: someone with a knowledge of computers will be able to pick up the method of using the package from the instruction manuals, but this is far from being the ideal situation for two reasons. First, there will usually be only a very limited number of such people on a project; their—or his, more likely—time may be too valuable to spend on this work. Secondly, if there is only one such person, the whole system is in jeopardy if he leaves the project, or falls ill. By far the best method of getting the package rapidly and effectively into full use is to send a small group of staff on a training course, and the provision of such training courses is an important part of software support. If this particular sort of support is not available, the instruction manuals do have to be really good: try them out on an intelligent but inexperienced user. If there is neither local training, nor first-rate, usable training documentation, then you really must consider either sending staff out of the country, or scrapping the whole idea of computerising the project management system: it is embarrassing to find you have spent money on software that doesn't work—but a lot less embarrassing than having spent it on software that works, but not for you!

The cost of getting enough staff adequately trained—so that there are stand-by personnel in case of sickness, and so on—must be considered before the decision to computerise is made.

(c) The program should run reasonably quickly—this isn't crucially important, with the project's own microcomputer, but it certainly is if you are paying for time on someone else's machine. Input/output rates are often fairly slow on many microcomputers—disappointingly so for users who have been used to large mainframe machines.

(d) The package should intercept error messages generated by the system software, and do something sensible with them; if, for example, the next item of input should be a duration, and you inadvertently start to key in the description of the next activity, somewhere in the machine a warning will be set off. This sort of error is, from the system's point of view, catastrophic: the program needs a number, which the system software must anticipate will be processed arithmetically—but it has been given some letters, and the operating system will usually be designed to assume that, because trying to do arithmetic on letters will fail, there is no point in proceeding. Every good system allows the programmer access to the error warnings; a good programmer will use this facility to prevent the system stopping the program and thereby probably destroying or currupting the data already put in. This means that a good piece of program (written in a high-level language) to input, say, activity duration would look like this:

```
500   DISPLAY 'KEY IN DURATION'
510   READ IN DURATION
520   CHECK ERROR CODE
530   IF ERROR CODE=0 GO TO 580
540   SOUND BUZZER
```

550 DISPLAY 'DURATION HAS TO BE A NUMBER—TRY AGAIN'
560 SET ERROR CODE TO 0
570 GO TO 510
580 ... (continue with program)

The numbers are program step numbers; the 'display' instruction puts up the text between the quote marks on the VDU screen. This type of programming prompts the user to correct his input as he goes along.

Other potentially catastrophic errors which the system will object to, and which have to be treated in similar ways, include attempts to specify such a large job that it will exceed the machine's capacity (e.g. by accidentally adding a zero to the end of the number of activities) and providing zero as the value for any variable that is to be used as a divisor (because the answer would be infinitely large, and therefore bigger than the largest number the machine can store).

(e) The package should also clean up the data, by looking for inconsistencies in the material. This is a general requirement for any sort of program; in the specific case of CPA, the cleaning up process would look for things like loops in the network (e.g. activity 37 follows 36, which follows 35, 29 and 49, which follows 36), and loose ends (i.e. a segment that doesn't link to anything at all at its downstream end). Networks are never designed in this way—but the material input to the computer may imply such errors.

A good technique in program writing is to prepare the program in main sections—input, calculation and output—with a pause during which the user is offered the chance of reviewing and modifying his input between the input and calculation phases.

One of the big disadvantages of large mainframe installations is that they often only offer batch processing: the user has to transfer his data and specifications to coding sheets, which are then converted to punched cards; he then checks these cards, and returns them to the computer room, and waits—possibly for most of the day—until the day's work is sorted and returned. At this point, he may discover that all that has been achieved is a small amount of abortive work, because of trivial errors in the input, since there is no way, under this sort of system, of prompting the user to correct errors during a run. On his own microcomputer, he would have been offered the opportunity to correct his input as he entered it.

(f) When errors are detected, the program should generate intelligible error messages—'the number you gave me is too big', not 'error number 511 D at line 9110'—especially if you have a choice of three manuals to look through for an explanation of the error number. One very large computer manufacturer is notorious for this sort of unintelligible error message!

(g) The program should be designed so that the user can break off during input, without losing the work already done; there are several reasons why this is necessary, including urgent interruptions.

(h) The form in which the input is requested by the program, from the user, should be a convenient and natural one. People selling computers and software lay great stress on their power and versatility; given that, there can be no justification for the programmer asking the user to spend time and effort sorting his data into a form convenient to the programmer. In the CPA context, an unreasonable form of input would ask the user to feed in the numbers of both the preceding and following activities, since, given a complete list of activities, each with its list of following activities, the lists of preceding activities can be worked out by the program.

In general, because computers are very good at sorting and indexing, programs should be designed to accept input, keyed in by the user, in the order in which it is most efficiently recorded on working papers, with no transcription, coding, etc.

9.3.2. Specific
There are a number of requirements, not all met by every software package, which are specific to CPA. These are:

(a) The package should be capable of producing bar-charts and tabular critical path analyses, for a sufficient number of activities. (Being able to get a printed CPA and network at regular intervals means that, in the event of a failure of the computing system, it will be possible to continue with hand calculations.) The ability to produce partial bar-charts, as instruction material for junior staff (Section 8.2) is extremely useful.

(b) It must provide facilities for resource scheduling which will both handle a reasonable number of activities and process them in a realistic manner; in particular, if only one method of scheduling is available, it should *not* use the day-at-a-time method (See Section 4.8). Graphic as well as tabular output is valuable, as it indicates clearly the periods when problems with particular resources are to be expected. Resource scheduling is probably the most demanding aspect of CPA calculations, and a heavy load of this sort of work is a powerful indication in favour of computerisation. Programs capable of resource scheduling typically require a microcomputer with a relatively large memory.

(c) The package should be able to handle crashing.

(d) It is essential that the network can be corrected or altered easily as, in practice, alterations to the logic of the network are quite common. Usually, the features of the program which allow this will also be used for updating, which is equally important: not only durations, but costs and resource changes must be easily revised or updated.

(e) The program should handle financial control well, producing both cost-prediction charts, and actual expenditure records.

(f) It should be capable of producing edited reports, so that the output can be produced, not only in the full form required at project manager level, but also in the format required by any officially imposed reporting system. However, it is not reasonable to expect that the package can produce the condensed reports needed for successively higher levels of responsibility in the sponsor's organisation: these require the exercise of thought, and of responsibility (see Section 8.3.3) and computers are there to relieve you of clerical work, not do your thinking for you.

(g) It is useful—but not essential—to be able to produce directly, on the printer, progress charts such as that shown in Fig. 8.6.

9.4. REQUIREMENTS OF A SUITABLE MACHINE

The point has already been made that, in general, you choose the software first; that will limit your choice of machine/system software combinations, and the choice among the latter should be determined by the following criteria:

(a) More important than any other consideration is that the make of machine chosen should be well supportèd locally, with spares and service staff readily available, and satisfied customers in evidence. If this criterion cannot be met, regardless of what other features of the situation are in favour of computerising the project management system, it is wiser to rely on hand computation. It is also risky buying a new model from an established hardware firm: if you have the first one in the country, you may well find the service engineers learning—slowly and expensively—how to service the new model on your particular machine!

(b) It should be available in an appropriate configuration, i.e. with the right combination of memory size, number of disk drives, and so on. (For example, the sample software described in Section 9.5 will run on various computers, needing from 64 K to 256 K of memory. Most CPA packages require two (floppy) disk drives, printer, VDU, keyboard, and a central processor with a relatively large memory.)

These two points are by far the most important considerations, as, despite sales claims to the contrary, there is not a lot of difference in machine speeds and efficiencies between the major makes of microcomputer. You do not need to concern yourself about things like wordlengths, or operating systems: if the software you want runs on the particular configuration of the machine you can get, these are automatically compatible.

9.5. EXAMPLES OF SUITABLE SOFTWARE

Two examples of suitable software will be described, one from the simpler end of the scale, the other, a very sophisticated package; both are produced by UK software houses.

(a) Pertmaster.[1] This is produced by the software house Abtex. Its minimum memory requirement is from 56 K (on a machine using the CP/M operating system) to 128 K (with the MS/DOS operating system), provided that two disk drives are available; but this minimum goes up to over 250 K if only one disk drive is available. (This underlines the importance of getting the right configuration of machine.) It also needs a VDU and a printer with a 132 character line width.

The maximum number of activities that can be handled varies from 700 to 1500, depending on the configuration of the computer—but even 700 activities represents a large network; the package gives the user the choice of activity-on-arrow or activity-on-node formats. In the latter case (referred to by the authors as 'precedence' notation), the lag form of notation (Section 2.10.2) is also permitted. Activities may be designated as 'milestones' (in the sense of major markers of the project's progress), so that condensed reports can be produced, and there are other facilities for selective reporting. Activities can be given pre-scheduled start and finish dates.

The reports' formats available include tabulations of the CPA: bar-charts showing activities, critical activities and float; tabulations and histograms of resource use, including cash. Lists of activities and their interlinkages are available, but not a printout of the network. All these items may be produced selectively.

Updating is achieved by altering the duration of activities to reflect the amount of work still to be done, with completed activities being assigned a duration of zero. The network can be readily corrected. Conversion to real calendar dates is possible and there are numerous other features.

Unfortunately, there is no method of crashing or resource scheduling—these have to be done by the user manipulating the other features of the program—and detection of loose ends and loops in the network is only possible at the analysis stage.

(b) Hornet.[2] This is a more sophisticated package, but you pay for the additional facilities, not only directly, but also through the more expensive minimum configuration of the machine. This package is produced by Claremont Controls.

Hornet requires a 256 K machine using the MS/DOS operating system with two disk drives, and a 132 character-wide printer. Only precedence notation (activity-on-node) is acceptable.

A large number of activities can be handled: a master network can be set up to control up to 255 sub-networks, each of up to 255 activities. Lag notation is accepted, and pre-scheduled dates may be set for activities.

In addition to a wide range of tabular reports, bar-charts, and histograms, all of which may cover all activities, or only selected groups, the powerful report generator program allows extensive modification of the output format, and an experienced user could probably get many standard official reports produced directly by the package.

Updating and calendar conversion features are good, and easy to operate.

Resource scheduling—on up to 8 resources per activity—is possible; the method of allocating a particular resource to the competing activities uses, among other things, float, and a user-assigned priority code, which is probably a satisfactory alternative to the criteria described in Section 4. Crashing is accomplished—somewhat indirectly—by assigning maximum and minimum duration limits to activities. There are additional features not described here.

9.5.1. Computerisation and Statistical Considerations

Neither of the packages described deals with the stochastic aspects of activities, described in Chapter 6. There are two reasons for this.

The first—which certainly influenced the originators of Hornet—is an even deeper distrust of the assumptions required to establish the validity of the methods (see Section 6.3.2) than that felt by this author. The second reflects a more general feature of computing: with a computer, it may often be easier to vary the whole problem, and re-solve it, to explore the effects of different sets of assumptions about the true values of the input data, and their true relationships, than to attempt to produce a definitive answer that embraces all possible outcomes. In the CPA context, this means that variations of the problem—representing likely variations of important features of the network—would all be solved, and the results used by the project manager to gain a feel for the likely range of outcomes. He would not have probability distributions for the dates of imporant events, or criticality indices, but he would probably be no worse off for that.

REFERENCES

1. Abtex (1983). *Pertmaster Users' Guide*, Abtex, Aberdeen.
2. Claremont Controls Ltd. (1983). *Hornet Project Management Systems for Microcomputers*, Claremont Controls Ltd, Rothbury.

Index